Q 175.52 .U5 M85 1989
Mukerji, Chandra.
A fragile power

W9-BSD-233

STOCKTON STATE COLLEGE LIBRARY
POMONA, NEW JERSEY 08240

A Fragile Power

A Fragile Power

SCIENTISTS AND THE STATE

Chandra Mukerji

PRINCETON UNIVERSITY PRESS

PRINCETON, NEW JERSEY

Copyright © 1989 by Princeton University Press
Published by Princeton University Press, 41 William Street,
Princeton, New Jersey 08540
In the United Kingdom: Princeton University Press, Oxford

All Rights Reserved

Library of Congress Cataloging-in-Publication Data

Mukerji, Chandra.
A fragile power: scientists and the state / Chandra Mukerji
p. cm.
Bibliography: p.
Includes index.
ISBN 0-691-08538-2 (alk. paper)
1. Science—Social aspects—United States. 2. Science and state
—United States. 3. Research—United States—Finance. 4. Federal
aid to research—United States. I. Title. II. Title: Scientists
and the state.
Q175.52. U5M85 1989
305.9′5—dc20   89-32314

This book has been composed in Linotron Caledonia

Clothbound editions of Princeton University books
are printed on acid-free paper, and binding materials are
chosen for strength and durability. Paperbacks, although satisfactory
for personal collections, are not usually suitable for library rebinding

Printed in the United States of America by Princeton University Press,
Princeton, New Jersey

10 9 8 7 6 5 4 3 2 1

To all my parents and my children

Contents

List of Illustrations

Preface

THIS IS NOT the book I started out to write. When I first went to Scripps Institution of Oceanography to do some field work, I intended to study technological innovation and science—not on a grand historical scale, but in the work world of laboratories. I still sometimes wish I had written that book; it would fit better in the context of contemporary science studies. But, although many of those I interviewed embraced my project, I also kept encountering others who gave me quizzical looks, questioning my focus on machines and technique. I eventually learned that "the look" expressed in large part a prejudice against technological innovators in science that I might well have ignored, sticking to my original purposes. But at the time, I did not know this. Moreover, I was curious to find out what *scientists* thought was important, if it was not technique. I found out, predictably enough, that they cared most about their scientific identities, the character of their research, and its reception among their peers. They cared about the kinds of issues that had preoccupied researchers in the Mertonian school of science studies. Second to this, they seemed more absorbed with the frustrations of trying to get and administer funds to sustain the research on which their reputations were based. Once I understood the latter, which was not at all well researched in the literature, and I saw how much of the life of the laboratory was devoted to generating or justifying research funds, I started to ask more questions about money.

The shocking thing I learned first, which changed the focus of this book, was that the nature of the government funding system for science, in spite of its unique form in America and its centrality to American science, is something that scientists and science policy experts alike have difficulty defining (as an ideal) and describing (as an ongoing process). They can tell you where some kinds of research will get funds, and where people in other specialties can go; they can tell you a lot about how funds are allocated, and how to maximize chances of getting money for a proposed project; they can tell who gets more money than others in their departments; and some can tell you up-to-the-minute details of the budget battles that will shape future allocations for science.

At the same time, they can legitimate the funding system itself with lots of good arguments, and criticize it with just as many others. Clearly, getting government money for research is a political process, so the political knowledge and ideological positions about it abound. Advocates say things like science is good for the economy, and that a strong science and technology base is essential for a strong defense. Critics ask what the country is getting for its "investment in science." They see few signs of a contribution by scientists to the well-being of the country. Interestingly, no one asks what scientists give up to get government funds. That is beyond the customs of the community.

No one asks about the drawbacks of the system to scientists because, at the same time that funding is justified politically, research results are treated as unbiased findings. Scientists who can get funding can make intellectual advances. Good projects are the ones that get funded because peer reviewers, be they internal or external reviewers, can tell the difference between good and bad projects, and they let higher administrators in the agencies know how best to spend their budgets. Some of the time, practical concerns might enter into the ranking of projects for funding, but this is the exception rather than the rule. The rule is that good science wins money. Since the purpose of agencies is to support good science, the only important measure of the success of the system is whether or not Americans are doing good science. For the most part they are, so the funding system has justified itself.

The relationship between good science and America's future is *assumed*, not analyzed, just as the value of the government funding system to the scientific community is *assumed*, not studied. In other words, the nature of the relationship between the government (a political institution) and the scientific community (an intellectual institution) is obscured, covered with clichéd political slogans and forgotten because both advocates and critics are afraid of what they will find, if they look at it closely.

I looked at it very closely, and what I found was not so fightening. What I found was a loss of historical memory. Americans have forgotten why they funded science to begin with, and have not paid attention to the forces that have distorted their memories. Good science is not exchanged for funds; neither do scientists merely follow the political winds to scrape together the monies they need for research. Scientists are paid to develop expertise that they can acquire by doing good science or use to create good science. The government assures itself of advisors and technical experts to help develop policies vis-à-vis natural resources and build weapons systems. This exchange of expertise for money, which is set up to allow scientists to serve both science and the state, is in fact a very subtle system that is built on an even more complex system of communications, and they both have helped to define a balance of power between scientists and the state that is highly nuanced.

The purpose of this book as it now stands is to explain in detail the characteristics of this system, and to show where they come from. It is a book that many scientists may find offensive because it is designed to attack precisely those areas of science and science policy that polite people have learned to ignore. But, as my husband says, good manners and good social science rarely coincide, and with apologies to my grandmother, who loved good manners, I prefer good social science.

Although the analysis ends up as a critique of American science policy, I think of it as a *cultural* analysis of the relationship between scientific power and laboratory practice. For the power of science lies less in what scientists tell the government than in the cultural authority of science as an institution. It is this cultural authority that scientists feed when they make dramatic new discoveries, and it is this cultural authority (as much as practical help) that makes science so useful to the government. For the money it allocates to support research, the state gets the right and ability to use that authority to legitimate its actions and for continuation of the soft-money funding systems, scientists give up control of their cultural power.

La Jolla, California
March 1989

Acknowledgments

I WOULD LIKE to thank the scientists, technicians, students, staff, and librarians at Scripps Institution of Oceanography, Woods Hole Oceanographic Institution, Stanford, Harvard, MIT, Oregon State, and elsewhere who have been so encouraging of this research and generous with their time. Many of those who helped me most were scientists on whom I relied for interviews as well as advice. The norm of confidentiality requires that I not thank them by name, but I would like them to know how indebted I am in any case. I can and must thank Dru Binney for her help, particularly at the early stages of this project, when she arranged for me to interview scientists she knew. She was also a great help later on, when she made essential files available to me and my research assistant. Most of all, she sustained me when I needed encouragement, keeping me at this project even when I was not sure what I was learning or who would care if I learned something about oceanographers. I also depended on the good will and help of Deborah Day and her colleagues at the SIO Library, Larry Cruise in Documents at Central Library at the University of California at San Diego, and the archivists at Woods Hole Oceanographic.

Usually scholars name their colleagues as their greatest helpers on a project, but I must mention before them two nannies, Erin Houston and Terry Floro, who kept my children entertained as I took them with me around the country while I interviewed ocean scientists—first as a terribly pregnant mother of one, and then as a nursing mother of another. They helped turn my son into an aspiring scientist (which he may continue to be, if he does not read this book), and kept my daughter close enough that I did not have to sacrifice the pleasures of new motherhood for my career, or vice versa. I am deeply indebted. I should also thank my kids for putting up with the difficult parts of life-on-the-road-with-Mom. I hope some day Stephanie gets over her hatred of airplanes.

I also need to thank my most supportive colleague and husband, Bennet Berger. He read all the early drafts of the manuscript, and scared me by actually liking them. In addition, I am indebted to Chuck Nathanson for early support for the project, and his probing questions about the analysis, while I was first formulating it. I also want to thank Bruno Latour for encouraging my interest in technology and (on his intermittent visits to UCSD) making science studies so lively. And I am deeply indebted to Michael Schudson for asking hard questions and reminding me to write for my critics as well as my friends. Finally, I want to thank Aaron Cicourel for spending the better half of a day going over the manuscript with me, and also Bud Mehan, Tom Gieryn, and Leon Zamosc for their useful comments. I did not take all their opinions into account here, but I benefited from the advice nonetheless.

I should thank a series of graduate students who have worked on science-related projects with me, or helped with this study itself. First I must thank Jerzy Michalowicz, who was my research assistant. Then I want to thank Kathryn Henderson, Chantale Hetu, and Bill Brigham for learning science literature with me, and giving me useful comments on the manuscript.

Finally, I am indebted to Diane David, who transcribed all the tapes from the interviews onto the computer, facilitating this project no end. I am in awe of her patience, and grateful for her help.

As always, the virtues of the book result from a myriad of fruitful collaborations; the faults are entirely mine.

A Fragile Power

CHAPTER ONE

Scientists As an Elite Reserve Labor Force

IN THE SUMMER AND FALL of 1986, a group of marine biologists and geologists were notified by the Department of Energy that their funding for
studies related to deep ocean disposal of nuclear wastes had been cut out
of the next year's budget. The DOE had decided on a policy of land-
based disposal of wastes and no longer needed their services. This was a
moment of victory for the citizen groups that had been fighting against
using the ocean as a dump site; it was also a moment that revealed quite
dramatically the political character of research funded by the government
and the vulnerability of scientists to the political process. Suddenly, this
group of researchers, who had been the voice of authority in deciding
what should or should not be done in the ocean, were unable to buy
more paper to write on or pay telephone bills. Not only did they face no
prospect for renewed funding in the next year, but because the program
had been penciled out of the next budget without any provision for closing the administrative offices, some of the money from the current year
was being recalled for use in boxing papers and moving furniture.[1]

It is easy to dismiss these events as not particularly worthy of attention.
After all, these scientists were supporting a government policy that could
have had disastrous consequences; so perhaps their loss of funds was
something to be applauded and not intellectually questioned and dis-
sected. It is equally easy to think that these scientists were low-level consultants for the government, not "real" scientists, so their careers (or failures) could not really be worthy of serious consideration. But many of
these researchers were respected scientists in their fields, and they were
consultants on important policy issues, affecting the ocean and the future
of the nuclear industry in the United States. What they were doing with
DOE money to promote science and the state ought not to be dismissed
summarily. It ought instead to be of interest not only to those concerned
with the careers of scientists, but also to those concerned with the uses
of science by the government in the formation and legitimation of policy.

The budget cuts in this small corner of the scientific community were
not detrimental to the scientific community as a whole; nor was this kind
of treatment of scientists an unusual occurrence. Many similar episodes
had dotted the relationship between scientists and government funders.
Many more researchers had lost their jobs during the cutbacks of the

1970s. What makes this case interesting is the way it illuminates both the power of science and the limits of that power.[2]

In the modern industrial societies of the West since the nineteenth century and particularly in the twentieth century, scientists have increased their intellectual and social power. Today science is perhaps the dominant cultural institution in these societies, providing models of correct thinking for members of other social worlds.[3] Scientists also constitute a powerful social force by affecting the economic and military strength of nations. They can and do tip the balance of power among states, and they even may determine the security of the world population through participation in the nuclear arms race.

Yet for all its power, the institution of science is built on a fragile social base. Science gains much of its financing and most of its social power because of its usefulness to government, but science cannot easily prosper as an intellectual endeavor simply by serving the powerful on issues of their choosing. Scientists may be best able to shape policy this way, but, as a consequence, they also trade away their ability to determine the direction of scientific research. Scientists can maintain intellectual and social autonomy for themselves and for science only by addressing other scientists, not members of other institutions. This is a problem, since most researchers cannot afford to keep their labs going without outside funding. Conducting research necessarily ties scientists to nonscientists who provide funds. For research that has no commercial significance, this means dependence on government money.[4]

Conflicting needs to find funds (hence dependence) and to do scientific research that serves science (hence autonomy) have shaped the character of research in the United States since the Second World War. The state has provided much of the funding for science (reducing the autonomy of scientists), but it has also given researchers some autonomy by decentralizing the research programs that they fund. A large part of the scientific work done with "soft money" from the government is conducted in private labs or at universities, and attempts have been made to spread research around the country. This has certainly freed many scientists from centralized government control of research through a national laboratory, but it has not really diminished their dependence on the soft-money funding system.

The puzzle underlying the soft-money funding system is not why government supports scientific research at all. There are obvious reasons why the state wants to know, for example, what kinds of containers to use to house high-level nuclear wastes, or why they get odd sonar pictures that confuse military personnel in some parts of the ocean. The puzzle is why the U.S. government also supports in universities and private laboratories basic research done with no immediate or even obvious long-term benefit to the state.

To understand a bit better the government program of funding basic research and of using scientists as advisors on policy, I interviewed and observed scientists at a number of labs who were engaged in one or both of the following: research on deep sea disposal of nuclear wastes and research on hot springs on the seafloor. The vent or hot springs research was a prototypical piece of "pure" or basic research, one affecting basic ideas in marine biology and chemistry while also raising some questions in marine geology. The waste disposal work was a classic bit of applied research.

I found in both cases that government interest in research was not well explained by government need for the *information* generated by scientific projects. I was led to the conclusion that when government agencies dispense money to scientists for research, they do it not so much because the state is interested in maximizing its store of scientific information relevant to policy issues, but more because the government has interests in maintaining a *labor force* of skilled scientists available for consultation on policy issues. The communication between science and the government is less a text-based system for conveying facts and more a form of oral discourse for conveying opinions. Funding makes the expertise of scientists (the skills and knowledge they embody) more consistently relevant to state interests and visible to government agencies.[5] That is why the state funds scientists to begin with, and pays them enormous sums to do technologically complex feats of intellectual labor.

Take the example of the scientists working on seabed disposal of nuclear wastes. At first blush, the experience of these researchers might seem to contradict the thesis that scientists are supported for their expertise rather than their information. These scientists were asked to collect data on particular areas of the seafloor that were being considered as dump sites, trying to anticipate problems of dumping there. So the government clearly hired them to collect information. Right? Not exactly.

To evaluate the possibility of dumping, the government needed more than results of experiments. The DOE did not just ask the scientists participating in this program to send reports that described what they learned from their research; they asked them to go periodically to Sandia Laboratories (a government laboratory in New Mexico) to evaluate the results of the combined research efforts and to speak directly to the policy issues. The government, of course, wanted the information from the research, but it was more interested in the *expertise* of the scientists in evaluating the results of research projects. Scientists could draw on their broader education on, for example, sedimentation process, food chains, or the mobility of animals to weigh the policy implications of a dumping program or to suggest a new one. These scientists could also anticipate how their colleagues would criticize their research findings, if they were to question the dumping program and join with political groups trying to stop ocean disposal of wastes. The evaluative skills of the researchers were essential to making policy recommendations, and they were part of their expertise, not just the information generated from their research.

To some extent the research was designed to yield information, but the scientists did more than provide information. They trained themselves in a set of skills for setting up and monitoring a dump site, if such a site were established. They developed techniques and machines for gathering and analyzing samples relevant to radiation leaks at the potential sites. They trained themselves to think about potential problems, which they could then use to guide monitoring procedures. Both technically and conceptually, they made themselves skillful helpers of government policymakers, and announced themselves available for consultation on seabed disposal of waste in return for funds to continue their own research at DOE sites.

Scientists like these and their counterparts funded by the government constitute what I call an *elite reserve labor force*. Soft-money scientists constitute a reserve labor force in the sense that they are supported by governments and industries so their honed skills will be available when they are needed (by, for instance, the military in case of war, by industry in case there are major changes in the direction of the economy, or by the medical community if there is an outbreak of some new and threatening illness). A pool of scientists actively engaged in research will have, at least in theory, highly developed investigative and problem-solving skills that can be called upon to address immediately pressing problems.

What about scientists doing basic research for the National Science Foundation (NSF) or the National Institutes of Health (NIH)? Do they fit this model? They are, after all, paid by the government to pursue their own ideas, not address policy issues. They are certainly training themselves to develop the most refined research skills possible, but who in government is that training supposed to serve? Actually, these researchers have quite an important role: to define quality in scientific research. People in government need to know which scientific ideas and research technologies are respected by the scientific community because the advice given politicians based on research is only authoritative when it is ratified by this group.[6] Agencies such as the NSF help to define for scientists and politicians alike what constitutes quality research. NSF researchers outline the cutting edge in their fields and thereby also identify techniques or theories that are passé. They help to spread their work practices within the scientific community through their publications, and the government underscores the importance of their tastes through ongoing funding.

Agencies that support basic research also improve the labor force of scientists in another way: they insure that a minimum of researchers in different scientific disciplines and subdisciplines will be allowed to develop and exercise the highest level of research skills they know how to achieve. Thus in substantive as well as evaluative areas, NSF and NIH help train a skilled labor force of scientists.[7]

It may seem counterintuitive to think of scientists as a reserve labor force. When social scientists speak of reserve labor forces, they usually mean the pool of unskilled labor kept on welfare during slow periods in the economy and employed when the economy needs them. It creates some cognitive dissonance at first to put scientists in this category. They are clearly so busy with research and writing, so well paid, highly honored, and seemingly invulnerable to the political tenor of the times that they seem anything but counterparts to the welfare poor. But like the unemployed on welfare, scientists on research grants are kept off the streets and in good health because of the interests and investments by elites.[8] Scientists have long joked about the similarities of their position to that of Aid to Families with Dependent Children mothers, in a kind of uneasy recognition of their structural dependency on the state (and perhaps also to underscore the social inequities that make them so well paid by the government while the poor are being criticized for their dependency even while they get relatively little money). This analysis only elaborates these ideas more systematically and seriously.

Perhaps one can think even more usefully of scientists as a reserve force like the Army or Navy reserves, as a group of people kept in training by the government to sustain the skills the military would need in case of war. Scientists have been cognizant since World War II of their usefulness as a military reserve but have not examined carefully how basic research done for purely scientific purposes could fit government needs. What seems sociologically significant is that they are paid money to keep their research skills sharp on the agreement (for scientists, implicit) that they are "on call" to be mobilized when their services are needed. Members of government can ask them to apply their scientific expertise to practical problems, and the norm of reciprocity requires that they agree to do so.

Researchers who have teaching positions and who do their research as only part of their work lives come closest to a reserve force in the military mold. Their teaching and university service is equivalent to their civilian jobs. Their research, any development and use of equipment that interests the state, and their active consultation with funding agencies or other agencies of the state constitute their training and periodic mobilization by the state. They may not be mobilized in the same way as members of the military reserves because scientists are rarely asked to abandon their reseach labs entirely to do applied work for the government or even to act as consultants for government on a full-time basis, but they often feel some obligation to pay back their benefactors by consulting on a more limited basis. (They will attend conferences set up by agencies to discuss government policies, weapons systems, or other engineering projects; they will observe and evaluate existing applied programs; they will review proposals for NSF, NIH, or other agencies; or they will join commissions to study problems).

Equating scientists to a military reserve makes more sense when you realize that the United States began the soft-money funding system at the end of the Second World War. The government in Washington developed a new science policy for America that was shaped by two factors: the importance of scientists to the war effort, and the inability of the military to keep them on their payrolls in peacetime. Military leaders were loathe to give up their trained counselors and weapons developers who had had such strategic value, but there was also a strong distaste in other parts of government for the establishment of a system of national laboratories either inside or outside of the military. Scientists would have to be demobilized, but they did not have to abandon their pursuit of strategically valuable skills acquired during the war, if the government was willing to subsidize their research laboratories. They could continue their research on grants or contracts, while working in the university or private research institutions.

Scientists supported in this fashion were no longer active members of the armed forces but were still paid to keep working on the kinds of projects that would make them useful advisors, if needed. The point of the system was not just to reward science for its wartime value but to make its power available as a resource for the state, minimally as essential preparation for any future war. The Cold War managed to extend immediate postwar anxiety over military readiness, and helped to extend government support for science after World War II.[9]

Scientists were valued by the military in part because in wartime their work could constitute strategic assets or threats, depending on which government they served. Immigrant physicists had been at the heart of the Manhatten project and continued in the atomic research program after the war. If they had done their research elsewhere, American power would have been diminished. The same was true of other less visible scientists as well.[10]

Just as Piven and Cloward[11] argue that members of the underclass are kept on welfare as a reserve labor force, sustained because they are potentially politically volatile, one can argue that scientists are kept in funds in part to keep them from upsetting the balance of power. An unhappy scientist (particularly one whose work has military significance) may not be likely to start a revolution but may emigrate and affect political life by doing so. Most are not deemed important enough to carry much of a threat to the state, but scientists as a group and the most elite members of these fields have enough strategic importance (at least in the minds of the military) to be worthy of some special treatment and are kept on good terms with the government. That is why the scientific reserve force became so highly paid and many of its elite members became temporarily politically influential, particularly right after the war.[12]

Military interest in science in the postwar period has been complemented by interest in science as a source of economic innovation and growth, even though the latter will be emphasized less in the upcoming pages. The idea that scientific advances lead to economic advances has become commonplace, providing an alternative rationale for supporting the work of scientists, particularly so-called basic research. Government money has followed this logic and been used to foster basic research for just this reason, primarily through the NSF, but also through myriad small agencies like the National Oceanic and Atmospheric Administration or the National Marine Fisheries Service.[13]

What seems ironic about the position of modern academic scientists in the United States, given their value, is their vulnerability. Although as a group scientists may have value to the state and individual elite scientists may even be celebrities, the vast majority of soft-money researchers are not well known even in their own disciplines, much less in government. They are supported not so much for their scientific accomplishments or strategic value as for their success in finding a funding niche.

These scientists could be compared to the eighteenth-century writers who have been described as the kept pets of the aristocracy, needing to tell their patrons what they wanted to hear. The most critical readers of scientific activity today certainly present an image of scientific dependence that is not far from this one, but it is drawn in more somber tones. The consequences of scientific dependence are seen as much more ominous, since the issue at stake is not a change in the direction of poetry, but the control of nuclear arsenals.[14]

This image of dependence, while perhaps exaggerated, does help point to the vulnerabilities of most government-funded scientists that are real enough. While the truth is that soft-money scientists who get established as journeymen or women in their fields usually find the money to get by, most also have many proposed projects rejected and find their labs growing and shrinking according to the vagaries of their funds. Some PIs (principal investigators) even abandon research careers because the fight for funds becomes either futile or too frustrating. Many young scientists, for example, give up research careers before they are ever able to establish themselves as PIs.

There are two vulnerable groups of scientists who find their research lives most easily shaped by state (or private, for that matter) interests. New Ph.D.s are one. They often turn to industry for jobs, or find positions in national laboratories. In either case, their abilities to set their own research agendas are sharply circumscribed. If they are able to find soft money outside these institutions, they are then most likely to do the kinds of applied projects that have less appeal to more well-known researchers. The other group consists of more-established scientists who have had trouble keeping their research going. They suffer from the instability of their social station in part because they think they have earned the right to more stability. Like their younger counterparts, they often try to reduce their vulnerability in the soft-money funding system either by getting long-term contracts for "easy money" work (doing applied or routine basic science with little potential for intellectual growth) or they turn to industry for partial support. Some develop consulting relationships with industry, develop private businesses, or attract private benefactors who will endow private research organizations. These ties to the private sector reduce some of the pressure to derive salary and expense money from the government. But many scientists are simply neither entrepreneurial enough nor strategically enough placed to develop these safety nets, and they in particular share some (but certainly not all) of the vulnerability of the underclass on welfare.[15]

What helps to define the high status of scientists in this situation is partially that scientists cultivate highly valued skills, but it is also that, for the most part, scientists are able to maintain control over the *intellectual* results of their research. Certainly, they must file reports with the government describing findings they derive from research done with government funds, but often little is done with those reports. A researcher in the military I interviewed said the generals and admirals know little of the research done with military funds, even that done for explicitly applied purposes.

Scientists also know better than most government bureaucrats how to interpret their results and apply them to either scientific or practical problems. This means that government officials are partially dependent on them. The esoteric character of the knowledge system—the fact that science is rife with conflicts that outsiders find close to incomprehensible—combined with the emphasis on basic research in the funding system make the information gathered by scientists functionally more salient to science and scientists than to government officials and their political worries. This means that members of the government need scientists to interpret research findings, using their wealth of background knowledge.

Even when scientific findings are clear and salient enough to policy that government officials could conceivably read them, they may not have time to do so. Papers may come and go through their offices at such speed that they cannot develop even a coherent conception of what researchers in their programs are doing. Under these circumstances, images of the natural world (theories of how it works) are really left entirely to scientists. In this way, the process of doing research with government funds, even though it may be fundamentally motivated by government interest in empowering the state, contributes to an empowerment of science through its intellectual growth. This is how research funding from the government contributes to the intellectual successes of soft-money scientists, at the same time that it articulates their structural dependence on funding agencies.

The sense of control over their *intellectual* lives is perhaps what makes soft-money scientists so adamant in insisting that the government gives them money for research and leaves them almost entirely alone to pursue their work. They admit that they have to fill out lots of forms and file numerous reports to receive government money, but they do not see these activities as particularly beneficial to the state. They see themselves and their colleagues getting the vast majority of the benefits from this arrangement. Researchers can feel this way because on a certain level they are right. The government is not getting vital information from them; the soft-money scientists are getting vital information for their science from research paid for by federal agencies.

Soft-money researchers by definition do work in the university or independent research establishment in which their careers are measured by scientific achievements. Since funding helps them do science, they are clear beneficiaries of the funding system. Moreover, in the process of doing research they experience little if any supervision from "the government." Their funding agencies may want site visits or research reports, but this seems like bureaucratic file making rather than supervision. So they see themselves getting money and doing science. Period.

But government-funded scientists who feel this way tend not to notice their contributions to the state. They see it as an honor and means for exercising power (or just a bother) when they are called upon to sit on panels or review proposals for the NSF. They are thrilled or at worst indifferent if asked to join the National Academy of Sciences; they do not worry about how this institution consults with the state. They tend to be pleased when called to testify before a congressional committee or asked to serve on a commission to study this or that. Their mobilization by the state is hidden within the "reward structure" of science.[16] They cannot see their value to the state, so they tend to discount the significance of science to national policies. They worry about federal dollars for research and hope that the agency people can make a good argument to keep funding up, but they do not see themselves as having a political role, per se. They feel all the autonomy built into their relationship to the state while feeling so indebted to the state that they cannot see their structural vulnerability.

Both the dependence and autonomy of scientists have been aggravated since World War II by increased use of expensive research equipment. On one hand, technology has increased the power of science enormously. New kinds of equipment have enabled scientists to observe and study parts of the natural world that were simply inaccessible to them before (for example, a sensor sent by robot into an obscure part of the ocean).[17] Rockets and cameras controlled by computer systems have made scientists aware of new moons around the planets in our solar system; particle accelerators and cloud chambers have made physicists aware of a microworld of atomic particles; and new machines capable of rapidly reading DNA strings are promising to open up understanding of the inheritance of physical characteristics in living organisms. These instruments and many others have helped to bring areas of the natural world into the domain of science.

At the same time, machines have also tied scientists in new ways to the military and business. Researchers' capacities to understand the world have become dependent on access to information-gathering technologies, most of which are made by and for the military or business for nonscientific purposes. Additionally, in some fields scientists are now essentially barred from doing any research at all without outside funds because they need such elaborate and expensive equipment. Their successes in using new instruments to empower scientific thinking have contributed to their increased *dependence* on government, not increased autonomy from it.

The odd mixture of autonomy and dependence for scientists in this system is not particularly well understood by scientists, but it is often described by them quite nicely. A marine geologist shows his confusion and yet sophisticated "feel for" the structure in the following description of his laboratory and how it functions (obtained during an interview with a senior scientist at a major research institution):

> The way [Blue Laboratory] works is . . . take an organization like my laboratory, the [AGL], that's my own creation. I have to feed it. The institution charges me to live here. If I want to use the telephone I have to pay for every phone call, for a pencil or a piece of paper. I have to buy it from the company store. If I want to use a submarine I have to rent the submarine or rent the ship. I have my own crew of people, about twenty or thirty people . . . it's my job to feed them. . . .

. . . But the by-product of that is freedom, to do anything I wish and to take advantage of the setting here. . . . The problem with freedom is most people say they want to be free but when confronted with what real freedom is, most people don't want it. 'Cause real freedom is you're on your own and you're not being kept. And if you're not being kept, then no one cares 'cause you're not on their list. So then you don't exist. . . . And so real freedom can be very insecure. But some people respond to freedom sort of like a vacuum. If you say I'm gonna put you in a vacuum, some will say, well, I'm gonna suffocate and others say I can expand. . . . So freedom is the same way; it's what are you . . . how you look at freedom. You know, oh my God, I'm on my own or oh my God I'm on my own. And [Blue Lab] tends to be best for those that expand when placed in a vacuum. And so it's sort of a free-wheeling Darwinian place. But as a result, you know, you can do what you want to do, *if you can find someone to support it.* [Italics added]

Clearly, this researcher finds the greatest challenge of research in his freedom to set his own research agenda and strategies. The lack of direct control by the government (of the sort that might be found in government labs) is more salient to him than the fact that he must scramble like everyone else in his peer group for funds. His problem, as he sees it, is getting on "somebody's list," not avoiding government oversight. On the contrary, he wants the attention from funders that will provide him with money he needs for achieving greater public visibility. Doing what he wants to do does not seem to this scientist fundamentally at odds with finding money to support it. In his case, one reason is that what he wants to do fits nicely with what the government wants him to do. Some of what he does is applied work where the coincidence of interest is not difficult to locate. But he also does basic science that is relatively easy to fund because with it he cultivates intelligence-gathering skills and equipment desired by the Navy. He is free to do his work because the military wants him "on call" in case they need information from the deep ocean. And because he keeps his labor (his skills, time, and information) on reserve for government use, he is able to do science that empowers him as a researcher, and he provides science with new resources. His case is perhaps an extreme version of a situation shared by many government-funded scientists in a variety of fields.

How did some science come to gain this position in the United States? What has the United States government done to change the position of soft-money scientists through funds since the Second World War? How has the institution of science fared under this regime? How has the practice of science fit and not fit the aims of the funding structure in the United States? These questions are too complex to be answered definitively by any analyst, and they certainly are not modest enough to suit the modest research project I have undertaken, but they are the questions that have shaped this analysis.

The peculiar position of government-funded scientists in the United States is complex and unusual enough to deserve some contemplation by sociologists concerned with understanding the role of the state in modern societies. In addition, careful attention to the structure of soft-money research in the United States raises questions about what are appropriate ways of analyzing science and its practice within the country. Without some understanding of the structural vulnerability and power of scientists and their research, we will know little about the nature of science as an institution. And our analyses of both American culture and politics will be impoverished as a result.

It is conventional in analyzing the role of science in modern societies to study well-established physicists or biologists (i.e., elite members of traditional and prestigious sciences), and use them as models for the rest of science, even though they are exceptional cases. Instead I have studied members of laboratories (from all the social ranks in those labs) of one of the less prestigious sciences, ocean science, one whose status as a unified discipline as well as its status within the scientific community as a whole is questioned even by ocean scientists. Marine biologists do not like to be called oceanographers. Geophysicists who study the ocean often do not want to be confused with deep ocean geologists. More than the usual pettiness about self-definition and identification are at work because this area of science is so new that its institutional legitimacy is lower than in the traditional sciences. Also, the technologist for doing many kinds of ocean research have been so recently developed and their use so rudimentary that the practice of this kind of science can be quite clumsy (although sometimes also revolutionary in its consequences).

This might make ocean research seem too green, unusual, or unimportant to deserve extensive sociological analysis. But significantly, ocean research is also highly funded. Its politico-economic value is great enough for an established group of scientists to work on the ocean and be given very large budgets to do so. In 1983 oceanographers at universities and colleges received $124,259,000 for basic research, while physicists (with their greater numbers and prestige) received $260,947,000. In 1987 the gap was even smaller between the funds given by the NSF for basic research in physics and oceanography. Money for "Physics" and the "Material Research" allocations for research in physics was $136,774,211, while the money spent for "Ocean Science" was $133,745,105.[18] The surprising scale of support for ocean research is even more apparent if one recognizes that in 1968, 39.5 percent of the budget for ocean research came from military money not included in the figures for basic research, and in 1986 just under 40 percent of the Scripps Institution of Oceanography budget came from the Department of Defense (DOD).[19] Also, a great deal of ocean research is done in private institutions whose budgets are not entirely recorded in either of these figures, so the total budget spent on ocean research is even more imposing than even these data suggest. An NSF study showed that in 1986 there were 1,900 oceanographers in universities and colleges who received $275,524,000 for research and development from all funding sources. In contrast, there were 15,900 physicists and astronomers (unfortunately the categories were collapsed) receiving $718,631,000 from all sources. These figures highlight the surprisingly high per capita funding income for oceanographers, which reflects the dependence of these scientists on research funds. Oceanographers had $145,012 per capita compared to $45,196 for physicists and astronomers.[20]

With their funds, many of the scientists who do work in the ocean have contributed to major intellectual advances in science since the Second World War. The theory of plate tectonics, so central to the revolution in geology since the 1960s, was developed in part from information about the spreading centers at the bottom of the oceans. The seafloor was found to be spreading out from central (usually) mountain ranges, changing over geologic time the relationships among the earth's landmasses and the geological composition and shape of the seafloor itself. And more recently, work on hydrothermal vents in the deep ocean has contributed to chemical understanding of seawater, resolved some questions about plate tectonics, and brought to light an ecosystem that is unique on the planet, one fed by bacteria grown with geological rather than solar energy. So ocean scientists have recently become reputable voices in the scientific community as well as major users of government funds for scientific research.[21] But the major reason for studying them is that ocean researchers are more dependent than most scientists on soft money. There are very few salaried academic positions for oceanographers. Much more than traditional scientists, they live off their research monies.

> You can't understand the motivations and dynamics of the field unless you really understand the self motives, unless in fact you talk to people about, you know, where the money's coming from. 'Cause you can be talking to somebody who's ostensibly a professor who's getting only four months salary. . . . This is not true in any other field that I know of. . . . The reason there are such a few faculty positions in oceanography [when oceanographers] have so much government funding is that there's nobody to teach. It's not an undergraduate degree. I mean, it's almost an applied, you know, it's . . . it depends on what you mean by applied. It's an applied field. You take physics and chemistry and you apply it to the oceans. So it's a graduate field. Well, it's not like geology,

you know. Every university has a geology department. Oh you know, maybe one university has an oceanography department. [University A] doesn't do a very good job but it has a department. To try to justify the university-paid salaries! So the university is saying, well, oceanography is wonderful but it doesn't bring in fees. Or I can't tell their legislature that I need ten faculty to teach ten students a year, you see. So there are very few faculty positions in oceanography. In physics, of course, they're huge. Every university has a huge physics faculty. And then all the money in physics then goes into hardware and graduate students and post-docs and getting things done. Whereas the majority of the money [in oceanography] is used just keeping the PI off the street, the principal investigator off the street. And so the anxieties and distortions that go along with that are pervasive. . . . And so oceanography, for all its large amount of funding, doesn't get as large a bang for the buck as it should because the overwhelming proportion of the budget goes in salaries. So all of it's soft money. . . . The line research people at [Big Lab] are soft money except for the faculty there. And at [New Lab] and [Wet Lab], would you or I still call them schools of oceanography? All they have is eight faculty positions which they multiply by four. So if you're a professor of oceanography at [New Lab] that means you have 25 percent of your salary. You raise the rest. So these are all soft-money places, which means that there's incredible, irrational, subjective components to the so-called peer review system. Basically [they're] saying, why should we give him, why should he come back in and get more, and get money out of a pot that I'm depending on for myself? [He goes on to explain why a famous scientist who dropped out of oceanography would never be able to work in ocean research here again. Then he describes government labs in Europe, and returns to the United States. . . . In this country there's no constraint at all. I mean, if the oceanography budget doubled tomorrow, we'd double the next day as soon as you could get the proposals in. And then what happens when it shrinks? You call up a guy on the phone and say, well, too bad you're on the street; your proposal's on the floor. Tough. No, you keep guys going. That's what it's called. So the NSF has almost become, in oceanography, it's unique to oceanography as far as I know, has almost become an employment agency.[22] [Interview with a senior scientist at a major university]

It might seem that because of their greater dependency on outside funds, oceanographers are too much unlike other scientists to be a good subject for studying the soft-money funding system. But on the contrary, they provide an extreme case that exaggerates characteristics of the system that also apply to other scientists. They can be treated as an ideal type of the soft-money researcher in the same way that physicists have been treated as ideal-typical theoretically advanced scientists. There are obvious pitfalls to this kind of modeling, but the virtues are also quite apparent. Extreme cases highlight social patterns that might be visible but less easy to see in related cases. Oceanographers may be more in danger of losing their research careers entirely if they lose their funding, but other scientists may be equally devastated by a loss of funds. This does not mean that someone with a teaching position with a guaranteed salary could not keep some research going while doing the teaching, but many avenues of research that are very expensive would be eliminated by a loss of government revenue.

It is also essential to point out that oceanographers are not alone in their dependence on soft money. Physicists (especially those doing weapons-related research) have also been primary beneficiaries of the soft-money funding system and hence developed equal dependence on the system. Computer scientists have also flourished on government funds. This situation changed radically when the computer business soared and began to provide computer specialists with so many opportunities in the private sector, but until the 1970s, most of the Artificial Intelligence research, for example, was conducted with Office of Naval Research (ONR) money. DNA work has also recently encouraged strong ties between biologists and the private sector, but earlier, the absence of other sources of money for research made even nonmedical research biologists (with few defense ties) deeply dependent on the government. If they wanted to conduct research in a serious way, they had little other choice. So both sciences with strong ties to the Department of Defense and those with little strategic or commercial significance in their own ways created a population of researchers deeply dependent on government funding.

Ocean researchers may seem an odd group of scientists to study since they work in submarines as well as labs; after all, they go to sea much of the time and work in laboratories on board ship that cannot compare to the more sophisticated ones on shore. But in needing expensive and sophisticated equipment, they are quite representative of large numbers of researchers in the greater scientific community. They are like space researchers who are dependent on rockets to get themselves or their experiments into space, and physicists and astronomers who find themselves only able to do their work when they can get "onto the machines" (telescopes, particle accelerators, or whatever) that they need for their experiments or observations. For all these scientists, their need for funds is augmented by their need for expensive machine time, and their use of machine time is colored by their need for the prestige that will assure them continued access to this machinery.

Thus many of the problems of, strategies for, and consequences of living within a soft-money funding system experienced by ocean scientists hold for researchers in other disciplines. The culture of marine science simply projects an exaggerated vision of the vulnerability and power of science in its current social role.

To highlight the technological and fiscal dependence of the scientists I studied, I focused on deep ocean researchers. Oceanographers distinguish between blue-water and white-water researchers, those who work far offshore and those who work close to land, respectively. I interviewed exclusively blue-water scientists, but I was more selective than that. With a few exceptions, I talked to scientists who did work on or near the seafloor off the continental shelves. Just getting to their research sites was necessarily a difficult task; using equipment there to conduct experiments or collect samples was also a technological challenge. I also chose the two kinds of deep ocean research to follow that I mentioned at the beginning of the chapter: research related to seabed disposal of (primarily) nuclear wastes and the discovery of and research of hydrothermal vents. The former not only tapped contemporary practical interests of government, but also had ties to the radiation studies in ocean research begun at Bikini and continued with nuclear testing. Hence my research yielded some interesting historical insights into government funding, while also supplying information about how applied projects are developed and funded. In contrast, the work on hydrothermal vents was an extension of basic research on the midocean ridges that had been central to the theory of plate tectonics. Search for the vents grew out of geological understanding of the ocean floor. Research on hydrothermal vents was, in addition, raising theoretical questions in geochemistry and biology that seemed to be very provocative. Thus it promised to be theoretically important to the growth of marine science.[23] Some practical and theoretical justifications for funding highly technological research in the deep were covered by these two lines of research.

In the course of my research I conducted seventy-four interviews with sixty-three ocean researchers and/or science policy experts located in nine institutions around the country. I used snowball sampling to generate a list of the scientists, engineers, technicians, and graduate students to interview. Because of the research topics I emphasized, many of those I interviewed were in marine biology (18) or marine geology or geophysics (19). Many (21) had been centrally involved in the development and use of new research technologies, some sampling systems, some *in situ* measurement systems, and a number of imaging systems used for producing indirect observations (recordings) of deep ocean phenomena. I also talked to fourteen scientists who had been involved in science policy and/or worked at different levels of the major funding agencies that support oceanographic research.

Because of the difficulties of just getting to the deep ocean, including finding the funds to get there, I did not do any participant observation among these researchers while they were at sea. I did go on board a research vessel as it was being loaded for a cruise, and I was once asked if I wanted to go on an expedition, but I was not able to leave my family to go to sea. Instead, I found a substitute source of data. For a long period of time, videotapes had been automatically made on a research submarine, the *Alvin*, recording conversations between the two scientists on each dive, documenting some of their experiments, and recording communication between the sub and the mother ship. I selected for detailed study eight tapes made during dives on hydrothermal vents in 1982 and 1984, having viewed all the usable tapes from 1981 to 1985. I chose this small group of tapes because they were particularly full of information. On them the scientists were more verbal, the visual records of experiments clearer, or the number of successful experiments or attempts to collect data greater than normal. I also paid greater attention to tapes that documented dives scientists had described to me in the course of earlier interviews. In addition, I viewed some film records and listened to audio tapes from dives made in an earlier period before the videotape camera was on the submarine. I used these to compare them under descriptions of dives given in interviews.

I also used other kinds of archival sources to supplement the data I acquired in interviews. The ones I accumulated most readily and relied on most heavily were research proposals. Scientists seemed happy to give them to me, if only to get me out the door at the end of a long interview. I was not unhappy to exploit this situation, since the proposals spoke directly (albeit not always candidly) about the funding strategies used by scientists and their sponsors. I also acquired some correspondence from scientists' files and read letters in the archives of oceanographic institutions. They gave me more insights into the development of research projects, the role of equipment in research, and funding patterns for science because they often lacked the formality of proposals and provided more details than the memories of those I interviewed. Records of research laboratories were also in the archives. They sometimes supplemented interview data about the development of new instruments or ships. In addition, they helped give me a more historical view of funding patterns in ocean research.

But the primary source of information for the study came from interviews that I conducted with a tape recorder and in person (with occasional telephone supplements). Most of these interviews lasted for an hour or more; many lasted two. Some were followed with second and third interviews; one poor man was interviewed numerous times, answered myriad questions over the phone, and even gave me a guest lecture. In my questions I tried to follow the line of questioning most appropriate for the subject involved. In many cases, I wanted to understand how they developed and deployed equipment to do their research. In many cases, I wanted to see how their work related to other work on seabed disposal of wastes or on hydrothermal vents. I tried to elicit details of the research process in my questioning as well as information about funding patterns. I wanted to find out what made this work seem valuable to the scientists; what they would tell an outsider (like me) about its value; and how they translated its conceptual value into its funding value. I also wanted to know details about the use of technology, since I found technology a major reason for the recent flowering of deep ocean research as well as an essential link between science and funders, defining their mutual dependence. So I pushed in these directions until I reached a sense of clarity about how the activities of the scientist were organized and executed. I did this with varying degrees of success. Some scientists gave me formulaic explanations of their activities that seemed to have been developed long ago and reused for many audiences on many other occasions. Others were simply reluctant to tell me anything. They did not quite understand what I was doing and were not interested in jumping into my research under those circumstances. But the vast majority of those I studied took the project at face value and told me whatever I wanted to know about in a fairly ingenuous fashion.

In reporting my findings here I use pseudonyms to describe oceanographic institutions, laboratories, and individual scientists. I also describe the social characteristics of those I quote in relatively general terms. I have tried my best to make the voices in the text ring true without embarrassing those who gave me their time and information. I know many of them would have been just as happy to have their names used in the book, but in the interest of confidentiality for those who felt otherwise, I have been as discreet as possible. This means that I have had to edit some quotes (as marked) and have dropped references to particular scientists when it seemed prudent to do so.

How government-sponsored research was defined at the end of the war, how it has been modified since, and (most importantly) how it shapes the character of contemporary soft-money science (to the extent that it can be read through the experience of deep ocean researchers) are what will be addressed in the remainder of this book. The second chapter provides historical background explaining how science became salient to governments through industrialization in the nineteenth century. The third addresses how the two world wars at the beginning of the twentieth century recruited the U.S. government into the support of science, and how this shaped the careers of scientists, the growth of labs, instrumentation, and the funding system immediately after World War II. The fourth chapter contains data on the funding system, and how it is structured to shape the expertise of soft-money scientists. The fifth looks at the meaning of intellectual autonomy to scientists, the limits of their autonomy, their struggles to maximize it, and how their striving for autonomy makes them more trusted advisors to the state. In the sixth chapter, by tracing the origins of the machines used in research, I show how instruments create a kind of conceptual dependence of scientists on those in government and industry who make decisions about technological innovation. In the seventh chapter I look at the centrality but relative invisibility of laboratory technology and technicians. I discuss the importance of technique in the development of laboratory "signatures," and contrast it with the strong distaste many scientists express for technological tinkering and the marginality of technicians to the social world of science. In the eighth chapter I look at how scientists use the research process to empower science and scientists in spite of their fundamental dependence on the state. They do this through an alienation of nature in the research process. In the ninth chapter, I look at why the state needs the expertise of scientists (and not just their research findings) to serve state policy goals. I show how the territorial struggles that are central to the world of science make it a mine field of claims that only scientists can easily unravel and apply to practical problems, so their expert opinions have a political value beyond just the value of their scientific findings. Finally, in the last chapter, I consider the political consequences of the dual communication system in science: the use of published research results for the intellectual growth of science and the use of the scientific voice for the empowerment of the state. I show how scientists (through their advice) have provided the authoritative voice of scientific dispassion as an instrument for politicians in shaping policy. Scientists, by participating in the elite reserve labor force set up by the government after the war, have traded the politically powerful voice of science for a not-so-steady supply of funds and relatively great control over the intellectual life of science. Through careful boundary maintenance and control of research results, scientists have made scientific knowledge autonomous from state power, and made their research experience vitally important to the state. As a consequence, they have been given enormous budgets to do lots of interesting projects. They have consulted on important policy issues, and their voices have been potent influences on decision making. The problem is that they have also given the voice of science, the cultural power of their institution, to the government. American science, for all its touted power, does not control its own best resource.

CHAPTER TWO

The Development of State Interest in Science in the Nineteenth Century

THE CONTEMPORARY relationship between science and the state was established in the United States at the end of the Second World War, but it had its roots in the nineteenth century, when the state began to provide fiscal support for scientific activities. Science became tied in more complex ways to the world of business because of its usefulness in manufacture and trade, and this led business to advocate government-supported scientific research in areas that would stimulate economic development. The technological innovations made in industry, in turn, made possible the development of scientific instrumentation that produced new research results, and the resulting scientific success (stimulating heightened state interest in science among European countries) led to new interest in science on the part of the American government. During the Civil War, as well, science was significant enough in affecting the outcome of the war to attract military interest. For all these reasons, government attention to the activities of scientists increased.[1]

Intellectual and organizational reasons combined in the nineteenth century to make this a particularly crucial period for the earth and life sciences, in part because of Darwin's monumental achievement, which provided a model of how to use world-scale biological information for theory building, and in part because of the colonial system and industrial development, which gave Westerners new reasons to want to understand and control nature.[2] There were some related practical reasons why these sciences would have matured at this point. In some cases, the research technology needed to complete work did not exist at all until the massive technological tinkering of the late nineteenth century. For thought dependent on massive surveys like biology and geophysics, advances in transportation/communication and cultivation of support that made possible collection of huge numbers of observations were necessary prerequisites for advance. In addition, the need for better communication/transportation and knowledge of resources to enhance industrial expansion helped to spawn land surveys for the railroads, ocean surveys for laying underwater cables, and analyses of natural resources from oil reserves to stocks of fish.[3]

Ocean science in the nineteenth century was clearly one of the areas affected by the application of new technologies to research and the increased sponsorship of science by industrialists and the state. In spite of long-standing Western interest in the oceans, particularly the Atlantic, there was relatively little known about the benthic floor before the middle of the nineteenth century. Spotty readings taken with lead lines had established the variability of the ocean's bottom, but systematic surveys had not been made. This was partially due to technological problems; studying the deep ocean was (and is) necessarily difficult and expensive. Surveys even above ground were time-consuming and difficult enough, since they could not be very useful unless they provided great detail or covered large areas of uncharted land. These characteristics were magnified in deep ocean work, since direct evidence of the subject could not be used; all data had to be collected through instruments.[4] For all these reasons, there was little motivation for considering this type of work until the nineteenth century. Practical seamen and fishermen knew the most about the currents and topography of the deep, simply because they had some experience with them. But they had no real need to cultivate this kind of information, and neither did anyone else. There was no powerful call for a scientific vision of the deep. But interest in laying underwater telegraph cables and the development of new equipment for making these kinds of measurements suddenly changed all this.

THE TRANSATLANTIC CABLE

The story of the transatlantic cable is one in which we can see how the interests of entrepreneurs, recruitment of state support, and the development of new technologies combined to provide scientists with new understanding of the deep ocean. Moreover, it helps to show how the birth of oceanography was tied to these external, social factors. The ocean became available for science as it became interesting to others and as the means for communicating with the deep were improved.

The idea of laying cables underwater was actively pursued in the 1830s, when the river Hoogli near Calcutta was spanned for the first time. And by 1840, seafloor cables were already being planned. In that year Professor Charles Wheatstone exhibited his plan for a Dover to Calais cable. This exhibition included both his design for a cable and his carefully drawn maps of the floor of the channel.[5]

His maps were an interesting part of the exhibit. They tied the development of cables to detailed bottom surveys, and they presented information from the surveys in a type of visual language that linked science to commerce. His were hardly the first maps to be made of a seabed; there was already a map of the bottom of the Mediterranean that was published early in the previous century by the Comte de Marsigli.[6] What made Wheatstone's maps interesting was the way he translated depth sounds into a cross section of the channel basin. This type of imagery was practically useful for planning cables routes, since it could be used to estimate the amount of cable needed to span a body of water, and it could be used to anticipate when there would be strains on the cable as it descended a precipice. It also was a form of imagery particularly well adapted to scientific investigation of the ocean floor.

Michael Lynch, in his research on biological slides, has noted that scientists like to transform bits of evidence into graphlike images. They will cut bits of a brain to highlight striation, thus making the brain look like a graph. They favor slicing nature so that it begins to resemble simple lines of measurement. It is a method of affirming the empiricism of scientists while also making nature more amenable to theoretical/mathematical analysis. In this kind of imagery the world of descriptive science meets the world of measurement in a single image. The cross-sectional views of seabeds may not have been examples as clean as the sections of brains studied by Lynch's subjects, but they do illustrate the same characteristics as scientific imagery. They are pictures that convert measurements into images and vice versa.[7]

The language of Wheatstone's maps, then, exhibits the marriage of practical and scientific interests in deep ocean surveys. The maps provide practical information for planning cable placement in a visual form that satisfies the interests of the scientist. They gave both scientists and entrepreneurs the means for envisioning and measuring what they could not see.

By 1843 even more cable plans were being projected. In that year Samuel Morse suggested to the U.S. government that one could lay a telegraph line even across the vast width of the Atlantic.[8] Thus by the early 1840s in the United States and Great Britain, where the Atlantic cable was first conceived and nurtured, the idea of deep water cables was already in the air.

State interest in the seafloor in this period seems to have been stimulated by the interests of entrepreneurs in laying ocean cables. Just in 1843, when plans for laying cables were being discussed in detail in both the United States and Britain, a series of voyages were undertaken by the British Navy to survey the ocean's floor.[9] And just before Cyrus Field began his serious efforts to lay a transatlantic cable, the U.S. Navy entered into the business of seafloor sounding, and provided exactly the kind of survey information that he needed to begin the work. The Navy, acting through the Naval Observatory, sent a Lt. Berryman across the North Atlantic with surveying equipment to study the ocean under the major sealanes. Some new survey equipment had just been built both for measuring depth and for bringing back small samples of the ocean's bottom, and the Navy apparently wanted to test the equipment.[10]

The survey by Lt. Berryman, although a serious effort at doing state-supported basic science, was not conducted entirely for nonutilitarian purposes. Lt. Berryman's report made explicit reference to a plateau he had "discovered" in the North Atlantic, which he dubbed the "telegraph plateau," precisely because he thought it was a good route for a cable. Thus a practical (i.e., politico-economic) vision of the seafloor was already alive during the time of Berryman's survey.[11]

When Cyrus Field, the entrepreneur, first entertained the idea of laying a cable between Newfoundland and Ireland, he wrote to two prominent scientists for advice. One was Samuel Morse, and the other was Matthew Maury. He wrote to Morse to find out if a cable of such great length under the ocean could in fact carry telegraphic messages, and he wrote to Maury to see if laying such a cable in the ocean was feasible.[12] Maury relayed information about the Berryman voyage to Field:

> This line of the deep-sea soundings [from Newfoundland to Ireland] seems to be decisive of the practicability of a submarine telegraph between the two continents, *in so far as the bottom of the deep sea is concerned*. From Newfoundland to Ireland, the distance between the nearest points is about sixteen hundred miles; and the bottom of the sea between the two places is a plateau, which seems to have been placed there especially for the purpose of holding the wires of a sub-marine telegraph, and keeping them out of harm's way. It is neither too deep nor too shallow; yet it is so deep that the wires but once landed, will remain for ever beyond the reach of vessels' anchors, icebergs, and drifts of any kind, and so shallow, that the wires may be readily lodged upon the bottom.[13]

Maury's optimism helped to convince Field that the project was worth pursuing, and Field's decision to lay the cable whetted his appetite for further study of the Atlantic seabed. Two further surveys were made by both the American and English navies, the latter indicating that the bottom was not as smooth as Berryman had described it, but also indicating that the ocean in the North Atlantic never reached the great depths that were known to exist farther to the south. This information helped to convince Field to stay with the northern route he had chosen, and to try to lay the telegraph line when he could secure the proper kind of cable.[14]

The laying of the transatlantic cable is primarily a story of how entrepreneurs and governments exercised their economic and political power to serve their interests. It is also a story of how the progress of science was furthered by the empirical needs and results of these practical actions. Learning about the ocean was not a monopoly of scientists in this situation; quite the opposite. But scientists used the opportunity to exercise their authority over the natural world. They acted as consultants to the military and businessmen; they also tested their theories on the ships and materials that governments and businessmen supplied. They turned the cable-laying project into their own experiment by convincing those with greater power than their own that their expertise was needed in case there were problems in laying the cable or transmitting through it.[15]

The power of the scientists in this project was limited by their ignorance and disagreements, but it was also enhanced by the results. They simply did not know at the beginning whether electrical impulses could be sent over such a long distance underwater and be clear enough to "read" at the other side. And they could not be sure that a cable could be safely laid on the ocean floor. They had to discover how much electricity to use to send a message through the cable, and they had to build new receivers to be sensitive enough to pick up the result. They also watched the play of the cable as it was let out behind the ships to evaluate their own surveys of the bottom, and they were forced to confront the difficulties of developing wires that were both appropriate for life on the bottom of the ocean and were light enough to be carried aboard ships to be played out over long distances. As the practical problems multiplied, so did the opportunities for scientists to learn from mistakes and try out their new ideas. And as the venture proved successful, they were in a position to profit scientifically and with new sponsorship from increased interest in cables.[16]

For Field and his people, success came slowly, but it came nonetheless. The first cable was laid across the Atlantic in the fourth attempt to complete this feat. The first voyage in 1857 ended when the cable started to be played out too rapidly. A brake was applied to slow it, and it was applied too fast, and the cable snapped. But before this accident, 335 miles of cable had been successfully laid and used to communicate with shore. This experience suggested that the project could, in the end, pay off. But it took a year to arrange the next attempts. Finally, in 1858, a cable was successfully laid the entire length of the Atlantic, but it mysteriously stopped working. Much discussion followed this disappointment to determine what could have gone wrong and how to prevent a problem in the future. The inadequacy of scientific understanding of both the sea and telegraphy were quite apparent. Yet scientists were more convinced than ever that the job could be done (and apparently were all the more interested in continuing the project). And in 1866, a functioning transatlantic cable was safely laid.[17]

The process of laying cable lines brought back much new data about the ocean that affected many areas of ocean research. For example, scientists not only learned about the shape of seabeds, but they also learned about the life found on them. Until the late nineteenth century, scientists accepted and reaffirmed the theory that high pressures and low temperatures below 400 fathoms made life at that or greater depths impossible. In spite of mounting contrary evidence and in part because of the numerous problems in sampling deep ocean temperature and fauna, this theory remained until the flurry of submarine cable laying. Sea urchins and starfish were routinely caught in the hemp lines used for surveys in the Mediterranean and Red seas before cables were laid there. And when the HMS *Bulldog* surveyed the North Atlantic in 1858 for Cyrus Field, the naturalist G. C. Wallich found thirteen starfish at 1,260 fathoms. This kind of information was ignored, since there was no clear evidence that the starfish were from the bottom. It was possible they attached themselves when the lines were going up or down through the water column. But when a broken cable from the Mediterranean in waters over 1,000 fathoms was retrieved for repairs, it was found with fifteen animals *encrusted* on it. Even skeptics had to admit that these animals could not have encrusted themselves on the line while it was being raised from the deep. Thus it was taken as crucial evidence for rejecting the old theory of an azoic zone in the deep ocean. Of course, scientists wanted more evidence than this, but these provocative results were used by Wyville Thompson and William Carpenter as reasons to dredge the ocean floor, looking for life. And they found it in abundant variety.[18]

The Thompson-Carpenter discoveries not only changed the reigning images of the nature of the deep sea; they initiated a number of scientific controversies about the character of the ocean that stimulated new research and successive waves of theoretical debate. Scientists began to generate images of nature that were global and systemic, and were based on data that lay people could not acquire. This helped to make the ocean the province of science, one that scientists tried to understand and render both transparent and tractable to others.

The scientific takeover of the ocean began in earnest with a voyage that is generally described as the starting point for modern oceanography, the voyage of the *Challenger* from 1872 to 1875. The major purpose of the trip was to collect basic information about the topography of the seabed, evidence of life forms and inorganic materials found on the ocean bottom, the character of the ocean's circulation, and temperature differences in its waters. The expedition was not primarily meant to serve the practical interests of business and government, but it was deeply indebted to the new interest in laying undersea cables and the scientific findings that resulted from them. Interest in the shape of the seabed had obvious practical connection to cables. The search for life in the deepest ocean, which was a continuation of Wyville Thompson's work just mentioned, and William Carpenter's theory of ocean circulation, based on temperature differences, were more purely scientific projects but ones that gained importance from the newfound interest in the deep sea.

Historians of science, interestingly enough, in spite of their great interest in the ground-breaking character of this expedition, do not generally draw any connection between it and the laying of deep ocean cables. They do not appreciate how practical concerns gave scientists the opportunity to do their scientific work. The one historian to mention the connection, Margaret Deacon, only speaks of it to deny that cable laying had any influence on the voyage. Yet some newly discovered letters written home by a crew member on the *Challenger* illustrate the self-conscious connection that the scientists on the *Challenger* saw between the two. In one letter, this sailor enclosed a copy of an address Thompson gave to the crew of the *Challenger* around March 3, 1873. In it, Thompson says:

> One reason why our ancestors did nothing towards lifting the veil from the sea bottom was because it was thought that no object could be gained by so doing; and the difficulties in the way were deemed insurmountable. For it was thought, and with reason, that nothing living could exist at a greater depth than about 400 faths. . . . It is now less than 20 years ago that scientific men began to talk of a scheme of ocean telegraphs, whereby the continents of America and Europe might be placed in almost instantaneous communication. How this scheme has succeeded we all know, but in the outside it was resolved by all practical men that some knowledge of the nature of the sea bottom was imperative, to enable the cables to be laid with any degree of accuracy and safety. Sounds, therefore, were at once taken across that portion of the Atlantic where the cables were to be laid. The Americans, as well as other nations, commenced a series of Atlantic soundings, which have been continued more or less ever since. The first dredging in deep water with anything like success was obtained by the Americans; a clever Lieutenant in their Navy, invented an instrument similar to the one we now use in sounding, whereby a very small portion of the bottom could be brought up. Their dredging, however, was in comparatively shallow water, and it was reserved for an English expedition, in 1863, to bring deep sea dredging to something like the system we are now working on. . . . Today the dredge brought us up an immense Prawn, or Lobster, from 200 fathoms of water, some of you perhaps saw it, and noticed that it lacked eyes. Now a common Lobster has a pair of very bright eyes fixed to the end of a sort of twig, and you may have seen it occasionally sling its eyes over its shoulder and look behind. Not so our crab, where his eyes should have been, there was nothing; and the reason, I take it, why there was nothing, is, that as there is no light where the gentleman lives; no eyes he requires, as they would only be an encumbrance.[19]

In this speech, Thompson indicates how interest in cables, the desire to make soundings to survey the bottom for cable laying, and the search for life in the deep using dredges were all of a piece to him and his fellow scientists.

Conflating commercial, military, and scientific activities was also common among the sponsors of this voyage. For supplying the ship and readying it for the voyage, the British government wanted surveys for the Admiralty charts, and surveys of the deep ocean to assess the feasibility of future telegraph cables.[20]

The British were not alone in pursuing these kinds of practical and basic studies of the deep ocean. As Margaret Deacon says in describing the *Challenger* voyage:

> The year of 1875 saw the *Challenger* in the Pacific. But now she was not alone in the field. In 1874 the USS *Tuscadora* had made soundings in the North Pacific along a proposed telegraph cable route between America and Japan. This voyage had been more than a mere survey. [The Captain] tried new kinds of sounding machines. . . . He also made surface and subsurface current measurements.[21]

In addition, the Germans were in the Pacific that same year doing their own oceanographic research. Under direct government sponsorship, ocean research was enjoying a resurgence, and the results were starting to be used to define oceanography as a distinct and legitimate science.[22]

The voyage of the *Challenger* is routinely cited in histories of oceanography as the starting point for this young science, because this was the first ship outfitted specifically for research and sponsored primarily for providing empirical data about the world's oceans. The "purity" of its mission and the practice of science on board is what makes it seem such a watershed. But it can equally well be cited as an example of how practical interests of nations have been tied to the pursuit of "pure" science. The British government may have been somewhat altruistic or idealistic in financing this expedition, but British interest in controlling the world's oceans (and its far-reaching colonial empire) was at its height in this period, and the government had vital interests in finding out how to improve shipping, to increase British military advantage at sea, and to locate the best routes for laying out an instantaneous communications system around the world. Scientists were given equipment (the ship as well as what was on it) with which to pursue their own understanding of a corner of nature that had been made salient to political life. They knew what the education was for, but they also knew that the government wanted them to develop their own theories about the ocean, in case these turned out to be important as other practical problems arose. The important thing was that they were studying a subject that concerned the government, and were willing to do research under these conditions.[23]

British concern about sea power and transoceanic cables was obviously not the primary issue on the minds of scientists as they pored over the oozes and clays dredged from the deep while they were at sea for so long. It didn't have to be. It only had to be on their minds as they outfitted the ship for the voyage and took the soundings the government had requested. The rest of the time they could think about science. The legitimacy of ocean science was both ratified and expanded by sending this floating laboratory to sea. The results of the research helped to promote more systematic theorizing about the ocean, and studies of the samples brought back helped to draw scientists from all over the world together. The practical interests of the sponsor, the British government, in the end helped to make oceanography both more autonomous and legitimate as a science, while it also tied the new science to the practical interests of a government intent on consolidating its position as an industrial and colonial leader.

LABORATORIES AND THE INSTITUTIONALIZATION OF SCIENCE

Managing research in the maturing ocean science of the nineteenth century already required scientists to maintain a delicate balance between soliciting funds and maintaining their control of scientific thought. Part of what made oceanography come of age in the nineteenth century was the growth of laboratories exclusively devoted to research and sustained over long periods of time. The routinization of state funding for research and the struggle for scientific autonomy in the face of this funding both had roots in these institutions.

There was an important irony here. Ocean science gained some of its autonomy as well as a dependence on funds because of its value to the powerful. Scientists had to learn when and how to control and to communicate information to make more powerful elements of society dependent on their expertise. Accumulation of data was to science what accumulation of legitimate uses of violence was to governments and what the accumulation of capital was to modern business. Creating more esoteric scientific debates made science more powerful by preventing easy access to its secrets by outsiders. They allowed scientists to command worlds of knowledge that had enough potential practical use to be of interest to outsiders. Theoretically and sociologically, scientists used their intellectual autonomy to cultivate sponsorship and used their sponsored research to create their own patterns of theoretical development and institutional autonomy.[24]

Research laboratories, looked at as social institutions, function to organize interaction between groups of scientists and a variety of "significant others." They work to tap external sources of funds; they are used to organize the accumulation and manipulation of what Latour calls "traces" of the natural world (both samples and records of other manipulation of nature); and they are sources for the theories, publications, and reputations that are the intellectual currency of science. According to Latour, laboratories are centers of calculation because they are the place where all these "currencies" are exchanged.[25] They make science possible by bringing together the resources that are needed for research: equipment, skilled people, evidence of nature, and writings. They also give scientists a base from which they can assert and protect their interests.

That is one reason why the establishment of distinctive laboratories for oceanographic research, both the design and employment of research vessels and the designation of space to house samples brought back from the sea and the experiments of scientists on those samples, is so important. Ocean science could not mature as an intellectual venture until scientists had a distinctive political base from which to demand both attention and autonomy.[26]

These ideas help to explain the practical as well as symbolic importance of the *Challenger* mission to ocean research. The *Challenger* itself was outfitted as the first strictly oceanographic research vessel, and the *Challenger* Office, in spite of its funding problems and problems of political legitimacy, acted as a kind of "center of calculation" of the sort that Latour describes.[27]

> Wyville Thompson's achievement [at the *Challenger* Office] was to create a network of personal relations, which, in the absence of concrete recognition either academic or governmental served as the first step toward building up an international community of oceanographers. The immense interest generated by the *Challenger* expedition and the material brought back naturally drew scientists from many different countries to visit the team at the Edinburgh headquarters. . . . The visits and correspondence which were entailed in the preparation and editing of the reports served to develop bonds between the scientists and the *Challenger* team which in some cases were lifelong. In the space of a few years Edinburgh had emerged as the international centre for the marine sciences, and continued to enjoy this position until after [the *Challenger*] report was completed.[28]

Perhaps ironically, the success of this "laboratory" in giving ocean science a new power and autonomy created a number of problems for these scientists. The *Challenger* expedition had been funded in large part because of the politico-economic rivalry among Western nations and the desire by the British to make their nation dominant on the seas. Thus there was strong opposition to either hiring non-Britains in the *Challenger* Office or sharing data with foreign scientists. But of course the interests of *Challenger* scientists were best served by having the reports on their data written by the best possible scientists. The persistence of those in the *Challenger* Office in pursuing their vision of proper scientific practices, despite strong political pressure, and their ability to sustain the basic funding are part of the evidence of the maturity of this young science.[29] The seas were now becoming so much a province of science that governments needed these experts, and thus had to yield to some of their demands. The voice of science was made legitimate, then, not so much by the elegance of its thinking, but by the successful struggle by scientists to make their intellectual decisions more independent of either government or business interests. Strong intellectual entrepreneurs were needed, and they were found to do the job.[30]

The Growth of the Laboratory System in the United States

In the United States, early research offices were also set up within the government. At first they were established for strictly practical purposes, but as the century progressed and ocean research grew in importance as a science, these offices were increasingly devoted to theoretical activities. Finally, scientific work emerged from the government bureaucracy (at least to some extent) with the establishment of independent laboratories. At each step along the way, ocean research gained increased financial support along with institutional and intellectual autonomy.[31]

The U.S. Congress was the group that first decided that the government should fund ocean research to reduce loss of life and cargo to the seas. In 1807 it established the United States Coast Survey for this purpose, and in 1830 the U.S. Navy established the Depot of Charts and Instruments. Both these institutions were not originally intended to be oceanographic laboratories; they were expected to do the kind of practical surveying that would enhance navigation and trade. But both were transformed into quasi-laboratories by ambitious heads during the 1840s. Alexander Dallas Bache at the Survey and Matthew Fontaine Maury at the Depot each tried to make their organizations centers of ocean research for the United States, studying not only aspects of the ocean that directly affected navigation but also qualities of the ocean that were being debated in the world of science.[32]

Attempts to create an American ocean laboratory were matched by the effort to launch a major scientific expedition, the United States Exploring Expedition of 1838. A combination of national pride (Congress wanting to show that its scientists were as good as those in other countries) and economic self-interest (stemming from the assumption that such a voyage could bring back useful information for enhancing trade and fishing) helped to convince the government to launch this project. But the success of the voyage was hampered by problems of authority. The Navy wanted to be in charge of the expedition, since it was using Navy equipment and personnel, yet the Navy did not have its own scientists to do the work expected on the trip. The interests of science and government were again in conflict here as they were to be at the *Challenger* Office, but in the case of the United States, the authority of the scientists was severely limited before they could gain enough power to fight for their interests. Much of the material collected during the voyage was destroyed either before or when it reached Washington by ignorant bureaucrats who had no idea of what benefit the samples might be to scientists. Thus much of the data collected on the voyage could never be used by the scientists. And funding for analyzing the results of the expedition was both limited and curtailed before analysis was completed.[33]

The efforts to create self-sustaining laboratories both at the Survey and the Depot were more successful. As these offices were funded to continue routine surveys of the ocean waters, they were able to add to their normal duties projects that had more scientific legitimacy and interest. Bache began collecting sediments from the seafloor, and he also created a sustained study of the movement of the waters in the Gulf Stream. When Louis Agassiz and his friend Louis François de Poutales became involved in the Survey, they helped to make it a center for marine zoology and geology. Agassiz and Bache were in a particularly good position to do this. They were among the most famous of American scientists, both with international reputations. They were also among the core members of the elite group of scientists that came to be known as the Lazzaroni. Whether or not this group acted in a concerted fashion for political purposes to enhance the power of science in America, it certainly provided the kinds of connections for scientists that they could and sometimes did use to advantage. Thus Agassiz and Bache could use both their positions in the scientific community and the growing power of that community to make this government office into something approaching a true scientific laboratory.[34]

Maury, at the Navy's observatory, was less well connected and respected as a scientist than his competitors at the Survey, but his own ambition drove him to make that office also a bastion of scientific investigation. By 1849 he had been given a ship to conduct his own research. He began to be interested in deep ocean soundings, and started this research on his vessel's first cruise. By 1854 he had accumulated sounding data from his own research and from the notes made by other officers in the Navy, and used them to publish his *Bathymetrical Map of the North Atlantic Basin with Contour Lines Drawn at 1,000, 2,000, 3,000, and 4,000 Fathoms*. It was the first map of the entire ocean basin.[35]

Both Bache and Maury had what success they did as heads of scientific labs not because science had the blessing and interest of the government, but because they struggled to make the interest of funders in dominating the ocean serve their purpose as scientists to create a new level of mastery of the ocean through research. The government clearly needed scientists to conduct research; when they tried to put naval officers without scientific ambitions in charge of the expedition, they created a fiasco both for government and science. Thus when Bache and Maury turned their offices into laboratories, they could argue that their activities were vital to the state. At the same time, this could not be accepted without struggle. Later heads of these offices did not engage in the same kind of struggle, and thereby lost the position of these institutions as leading oceanographic laboratories.[36]

Just as this generation of scientific entrepreneurs was dying off and their institutional base for advancing science began to erode, a new group of scientific entrepreneurs began to establish independent laboratories using private funds. The idea for such laboratories had been nurtured by the earlier generation, particularly by Spencer F. Baird, who was in charge of the government's Fish Commission. He dreamed of starting the Marine Biological Laboratory at Woods Hole in Massachusetts as a center for serious study during the summers. But he had difficulty trying to obtain government support for this project, and he died before the institutions was really successfully established. In 1890, the MBL was successfully launched. In California, William Ritter began a summer program for biological research two years later, funded by the wealthy Scripps family. By 1903, the Marine Biological Association of San Diego was founded, and in 1912 it became the Scripps Institution for Biological Research. These new scientific entrepreneurs worked like their predecessors to make science a more powerful institution, but they also changed the institutional arrangements for research. State support of science was waning. In England, this was signaled by the loss of support for the *Challenger* Office. In the United States the autonomy of the scientists working for the government was increasingly questioned as they moved toward basic research. The whole democratic flavor of American science was based on the idea that science would have direct benefits for the people, not that an elite group of thinkers would get funds. So, much research moved into private hands, and some new kinds of research could flourish as never before. The growth of a new experimental biology at the Marine Biological Station is a case in point.[37]

The privatization of science in the United States contributed to a decentralization of research. The government, in its efforts to develop a democratic science, had already created some of this pattern, particularly by locating land-grant colleges and other applied-science institutions in different sites around the country. The growth of private research institutions and the development of serious research teams in industry continued this trend.

The privatization and decentralization of research in the late nineteenth century was associated with a "professionalization" of science. Scientists in the United States were more carefully distinguished from engineers, the practical men of the era, and many of them championed this differentiation because they saw it as an important maturation process for American science. Traditionally, they had been much more driven by practical concerns than their counterparts in Europe. With the development of new laboratories devoted to the pursuit of basic research, American scientists were able to compete successfully with Europeans and contribute to a worldwide spurt of intellectual growth in the sciences. Their primary social orientation was increasingly cosmopolitan and elitist, identifying themselves with the small group of pure researchers in the West. Their more applied counterparts, on the other hand, remained more nationalistic as well as pragmatic. Businessmen increasingly turned away from scientists in these private laboratories, nurturing more ties with engineers and the scientists remaining in government. Their disdain for pure scientists was matched by the scientists' contempt for applied work, and encouraged scientists to disavow any worldly interests in their work.[38]

FISHERY STUDIES

If one wanted to argue *against* a model of increased government support for oceanography in the late nineteenth and early twentieth centuries, one might look at the declining interest in physical oceanography after the return of the *Challenger* and the decline of government laboratories in the United States. Once the romance and the practical work of early voyages like the *Challenger* Expedition and the U.S. Exploring Expedition were over, there was not much public interest in supporting the ensuing large bills for data analysis. Scientists wanted most of all to study the samples brought back from these trips, so they fought to convince the government to supply them with a small research budget over a number of years. Still, they had to fight vigorously for the support because government desire to fund this kind of basic research was gone.

But declining research funds in deep ocean work were matched by growing research funds from governments for studying fish population and migration. Fishery studies were not and could not be effectively executed by private groups of fishermen. They required such large surveys of fish behavior that they were eventually organized into an international research program, but they began on a local or nationwide basis. There they provided an early example of ongoing government commitment to research based on problems of industrial development. They illustrated that the government could still be committed to funding science, when the research had clear national interests.

The fishery studies were clear examples of state-sponsored research to support industrialization. The mechanization of some kinds of fishing and canning made a steady supply of fish a commercial necessity. Problems in these industries began to draw attention to fluctuations in fish populations that were threatening commercial stability. Large-scale industries depend on routine. If they cannot predict their productivity, they are very difficult to run. Fishing does not fit well in this kind of system, since the number of fish caught often changes dramatically from season to season. Worst than that, declines in fish populations in the nineteenth century, when commercial fishing boats began using large nets to increase their hauls, raised suspicions (at least among line fishermen) that fishing areas were simply being overfished. That would have been a disaster for everyone—fishermen, canners, and local economies. But no one was quite sure whether this was the case. Were the fish just moving around or were they being seriously depleted?[39] Since regional and national economies would be threatened by a serious decline in fisheries, governments began investing in fishery research.

Since the 1860s there had been strong complaints about fluctuations in the fisheries, and these had led to the establishment of numerous national research teams for conducting research. But their efforts were hampered by the fact that fishery problems were not always local, and the reigning models of fish migration used to explain variations in fisheries were global in nature. Most practical seamen and scientists assumed that fish in the North Atlantic migrated annually from the cold Arctic waters down to the Atlantic fishing grounds. Usually they went to the same places each year, but sometimes they changed course and did not appear where expected. Local fishermen and scientists expected that the "missing" fish were appearing in large numbers elsewhere, but they were not in a position to find data to support or refute this idea.[40]

In spite of the difficulties in learning much about fisheries without access to more global data, much was accomplished in the late nineteenth century. For example, a number of early population studies were made, but they suffered from one important problem that needed immediate attention if scientists were to discover if and when fishing areas were being seriously overfished. They needed simple measures of the age of fish. A German, Friedrich Heincke, worked on this problem for years, and came up with the idea of studying fish scales. The uneven growth pattern of fish during the summer and winter leaves growth rings, not unlike the growth rings on trees.[41]

Johan Hjort of Norway then suggested that scientists could use scale analyses to study fish populations, using principles of vital statistics. Applying these methods he discovered that most of the fish caught by fishermen over a three- to five-year-period came from one particularly large "year class" of fish. He followed the 1904 year class until its final demise in 1911. Asking himself what caused these large year classes to develop, he began research that correlated large year classes with large amounts of plankton where the fish eggs were being hatched. The rich food supply for the newly hatched fish nourished greater numbers of these vulnerable youngsters.[42]

These lines of research on fisheries illustrate once again how the growth of marine biology and physical oceanography profited from the practical interests of business and from the state's acting on behalf of commercial interests during the industrial revolution. The culmination of this movement was the establishment of an international commission charged with studying marine issues. In 1899 King Oscar II of Sweden called together scientists from numerous Northern European countries (where interest in these issues was greatest), and asked them to establish guidelines for a massive research program. In a few years, the International Council for the Exploration of the Sea began to fund and publish a new regimen of research.[43] The shared problems of industrializing countries led them to cooperate with one another in making this and other large-scale research ventures work, and competition among these nations helped to spur national programs of research and development. Science both emerged from and tried to transcend the limits of the competitive' world economy. And in doing so, many sciences, including ocean sciences, began to establish a new legitimacy and patterns of growth.

As scientific research on the ocean became more extensive and institutionally solid, knowledge of fish behavior that had been almost exclusively controlled by fishermen was increasingly placed in the hands of scientists by governments. The needs of commercial fishermen and processors not only took away some of the business of traditional line fishermen, but also undermined their expertise, as designated scientists became outsider experts with a more global vision of the sea and fish. Once again the demands of industrial expansion gave scientists conceptual control over new areas of nature. And scientists used this "cultural capital" to increase their social as well as intellectual muscle.

SCIENCE AND GOVERNMENT AT THE TURN OF THE CENTURY

By the end of the century, then, both a set of local, national, and international laboratories funded by government and devoted to applied research and a set of private laboratories devoted to purifying science and advancing its cause existed side by side. In the United States the private laboratories were particularly powerful in making American science finally competitive with science practice in Europe, so they had a strong role in the social world of scientists. There were also fewer national laboratories with any intellectual charisma, so science in the United States became increasingly privatized in spirit as well as fact. Scientists learned to appreciate the resulting autonomy from government and from its pressures to make science useful. This does not mean, however, that scientists no longer cultivated practical skills. It only meant that many learned a kind of obliviousness toward the practical significance of their activities, which in the end helped to make them even more useful to the state. To the extent that scientists in this period learned to think of themselves as solvers of intellectual puzzles without regard to their practical meaning, they made themselves less likely to challenge state policies based on scientific research and hence even more attractive sources of advice.

Thus the movement of research initiative into private hands and the cultivation of basic science in this period helped to make the use of private scientists as consultants more attractive to the government, and to imbue scientists with the habit of working within a complex structure of private laboratories. In this way, the nineteenth century helped lay the basis for both government support of science and for the development of private science, which the government could use when the occasion arose. And unfortunately the wars of the early twentieth century provided many occasions for Western industrial countries to call on scientists to use their expertise to affect the international balance of power.[44]

CHAPTER THREE

War and State Funding in the Twentieth Century

JUST AS THE RELATIONSHIP between science and the state began to transform both corporations and laboratories in the nineteenth century, the
relationship between science and the military began to change the practice of war and research during the early twentieth century. In the period, war was an overwhelming historical fact for the West. The character
and ubiquity of modern warfare gave states new motivation for seeking
the advice of scientists. The outcome of war was increasingly tied to the
technological and economic strength of nations, and science could affect
both the technological sophistication of armies and the industrial power
of lands. Military interest in science immediately affected scientific
power and autonomy. As the military started supplying funds and new
technological media for scientific inquiry, scientists were given new
means for empowering science while also cultivating a new dependence
of researchers on the state.

OCEAN RESEARCH IN WORLD WAR I

Growing government interest in the military uses of science was already
evident during the First World War, although the changes were relatively modest at that time. The scientific bases for industrial development
(particularly in Germany) as well as the use of science to improve weapons and communication systems had by that time put science at the center of a new way of fighting. Chemistry was one scientific field with obvious strategic importance; it was vital for making the metals, explosives,
and dyestuffs needed to arm and clothe soldiers for war. By World War I
the Germans had outstripped other countries in the West in its chemical
industry so much so that most other countries had learned to depend on
German imports for the chemicals.[1]

Ocean science was also deemed strategically important by the military
by World War I, so the expertise and advice of scientists were sought by
the U.S. government. The submarine was the major reason. In the
United States in 1915, the Naval Consulting Board, headed by Thomas
Edison, set up a Committee on Submarine Detection by Sound to find
out what was known on the subject and what could be done to improve
the U.S. capacity to identify submarines underwater. The following year

the National Research Council (NRC) was established and set up its own committees, including one on submarine detection. Both of these committees helped to establish a precedent-setting relationship between marine scientists and the Navy, centered on the development of acoustical equipment to monitor sounds under the surface of the ocean.[2]

Most instruments developed for underwater acoustics before World War I were simply means for listening to underwater sounds, not creating and monitoring echoes. This kind of device could be and was used to listen for the engines of submarines, but it was a passive system. But in 1901, the Submarine Signal Company of Boston worked on a more interactive system. The people there developed receivers to listen for bells installed near dangerous rocks along the New England coast. Around the same time, some rudimentary forms of echo sounding (bouncing a signal off the seafloor) were also beginning to be studied.[3]

The sinking of the *Titanic* increased in developing echo-ranging equipment, that is, instruments which would give off a signal that would move horizontally through the ocean and bounce off of any impediment in its path. By 1915 this equipment was successfully tested as a means for locating icebergs. Reginald Fessenden, the man who designed the oscillator for this instrument, began to put his invention into service for the military in 1917, when he was called to consult with the Naval Consulting Board's Committee on Submarine Detection. The Navy was so taken with the device that it was already trying to adapt the machine to submarine detection when the United States entered the war. During the war the research continued at a Navy research facility in Nahant, Massachusetts. Simultaneously, private research in echo-sounding was being stimulated by the NRC's Committee on Submarine Detection. This group organized a laboratory composed of university physicists at the U.S. Naval Experiment Station in New London, Connecticut.[4]

The alliance with scientists was not so important to the military during this war as it would become in World War II. The effects on science were also not so great, since the amount of research funded during the First World War was small and after the war government funding essentially stopped. Still, scientists learned some lessons about acoustics that began to affect basic research.

For example, Dr. Harvey Hayes, one of the physicists brought to the New London laboratory, used his echo-sounding equipment (the Hayes Sonic Depth Finder) on the USS *Stewart* to make nine hundred soundings for the first continuous profile of an ocean basin. The fact that many of the soundings could be made while the ship was moving at full speed made the equipment quite remarkable to scientists who were still used to taking soundings by stopping the ship and tediously lowering and raising lead lines. In 1922 the USS *Hull* and USS *Corey* made soundings off the California coast, and the *Guide* made them down the East Coast and through the Panama Canal to the West.

In spite of the miraculous spread of the machinery, the techniques for using it to map the seafloor were not yet well developed. Checks of sonic depth readings against those made with lead lines indicated that the sonic measurements were distorted. It happened that the velocity of sound through seawater varied for different temperatures and salinities. This problem prompted researchers to develop tables to correct for these distortions and make measurements more reliable.[5]

OCEAN RESEARCH IN WORLD WAR II

The Second World War not only forged a more intense and complicated alliance between science and the military, establishing a contract system to support scientific research, but it also laid the basis for ongoing government funding. This system was perpetuated after the end of the war, albeit in a much modified form, mostly because science proved to be crucial to the outcome of the war. The atomic bomb may have most dramatically altered the wartime balance of power, but there were other, less heralded (or hated) innovations that proved vital. Developments in radar, echo sounding, and navigation contributed to a superiority in air and sea power by the Allies that helped to precipitate the defeat of Germany and Japan.[6]

The new relationship between science and government that developed during the war (and renewed in a revised form after it) was orchestrated by the National Defense Research Committee (NDRC) and the Office of Scientific Research and Development (OSRD). The NDRC was set up in 1940 to focus scientific attention on problems of the military. It set up a contract system for distributing funds for research to a wide array of institutions, from universities to the research laboratories of private corporations, taking advantage of the decentralized system of science established in the United States in the nineteenth century, but using money to direct the research programs of the various labs. The contract system helped to establish in the United States a policy of giving scientists some control over the design of their funded research, as long as the work served the national interest. The degree of autonomy given during wartime was quite small, for obvious reasons, but the idea of partial autonomy of researchers was established then. Money was used to keep scientific research on desired topics, and scientists learned to make the policy significance of their research clear to the government through "consultation and discussion":

The first requisite of a satisfactory organization of science for war is that it must attract first-rate scientists. One outstanding man will succeed where ten mediocrities will simply fumble. In creative thinking there are no substitutes for imagination and resources. These will flourish only when the scientist has ample funds and a large measure of freedom. At the same time he cannot work alone in an ivory tower. Many of the problems involved must be attacked by teams of men with different skills and angles of approach, and on all of them effective liaison must be provided with the armed services. The organizational problem at this point becomes one of great difficulty. Armies and navies are operated on the principle of the chain of command, not on the principle of consultation and discussion. Their systems are well adapted to the conduct of military operations and to the production and methodical improvement of standardized equipment. They are anything but favorable to the conditions under which scientific inquiry best thrives, and the reconciling of the two is a matter of the utmost importance.[7]

Scientists had been indicating since the nineteenth century that they needed autonomy to do their work, and were often resistant to serving government with their research. Henry Bigelow at Woods Hole Oceanographic Institution was quite plain about eschewing applied work for the military, referring to it as "plumbing." In the nineteenth-century mold, he treated applied research as the work of engineers rather than real scientists. But his attitudes did not prevent other scientists from finding contract research attractive enough to offset their prejudices. They lined up before the government's coffers. Many of them shared Bigelow's belief in science's superiority to and detachment from applied work, so they paid single-minded attention to the interesting intellectual puzzles within applied projects. They were able to sustain their images of science this way, but they also learned to serve the government while disregarding (to a large extent) the political and moral significance of their work. For these scientists, the structural autonomy designed by the government and the ratification by government officials of the scientists' own belief in the superiority of their disciplines made them better servants for the state, a lesson not lost to policymakers at the end of the war.[8]

Of course, many scientists, including oceanographers, knew perfectly well that scientists were given contracts to aid in the war effort, and they were happy to do their part. They did not find Bigelow's kind of elitism appropriate for wartime. Many ocean scientists readily went to work searching for technical means for improving naval warfare for the Allies. The extraordinary early success of German U-boats scared and fascinated them. They saw they could be useful to the Navy by figuring out how to disarm this menace, and by providing other kinds of advice about how to wage war at sea.

Oceanographers made their greatest contribution to the war effort in the area of echo sounding. The improvements in sonar equipment made it more useful to the Navy than it had been in the First World War. Directional echo sounding was used to scan the oceans to find out where enemy ships might be. New kinds of equipment such as improved hydrophones or display screens were developed to make sonar equipment even more sensitive and versatile.[9]

Oceanographers also solved a number of problems that the military encountered in putting this equipment to use. One was developing accurate and swift weapons to destroy submarines, once they were located. Depth charges were often slow in reaching their targets, so submarines could easily escape from them. In this way weapons research became a job for ocean scientists who knew something about the properties of water and objects moving through it.[10]

Another group of scientists began studying an early problem with echo sounding. It was what Columbus Iselin dubbed the "afternoon effect." Echo sounding became disrupted during warm afternoons when the water was relatively still. An investigation revealed that temperature changes in the water deflected sounds moving through the ocean. When near-surface waters were heated up by the sun (as they were routinely, particularly in the Pacific), creating distinctive layers of warmer and colder water, submarines could "hide" near the boundaries of these layers, where the sound waves would be deflected. Clearly, those either running submarines or interested in finding them had reason to want to locate the layers of water down to submarine depth.[11]

The bathythermograph (BT) was designed to do this. The idea of an instrument that could measure changes in temperature through the water column had existed for a long time and was partially realized by Carl-Gustav Rossby and Athelstan Spilhaus beginning in 1934. But in 1940 Maurice Ewing and Allyn Vine designed a version of the BT that was easier to use, and then used it to study sound propagation. They presented a report on the subject to the Navy, and soon after this the BT was adapted to shipboard use. It was placed on submarines by 1942. Throughout the war the readings made with these machines were accumulated, and they provided scientists of the postwar era detailed data on water temperature, particularly of the North Atlantic.[12]

Other researchers looked into the sound reflectivity of the seafloor and the characteristics of sea creatures that might affect what was heard with sonar devices. They learned that sound was reflected differently by soft sediments, a smooth hard bottom, and rocks on the ocean floor. Along much of the ocean floor, sound waves were reflected directly back to ships from the smooth, hard sediments of the seabed. But in other areas they would bounce around wildly off rocks or be absorbed by soft sedimentary layers, making it difficult to locate a submarine. Therefore maps of the seafloor became something of strategic interest. Similarly, biological studies of the sounds emitted by sea creatures became interesting to the Navy and to Martin Johnson at the University of California's Division of War Research in San Diego. Johnson began to study sea animals with a hydrophone to isolate the sources of certain noises. The most bothersome ones were made by snapping shrimp. They created such a loud and continuous crackling noise in some areas that sailors mistook them for some kind of jamming device invented by the Germans.[13]

In addition, surface studies, particularly studies of wave action and other varieties of water/air interaction that affected the functioning of ships (including weather patterns) were taken up by ocean researchers. Harald Sverdrup and Walter Munk made major advances in predicting wave activity and then used this information to advise the Navy about the feasibility of amphibious landings. They started simply by studying wave action in the open ocean. There they determined that the shape of waves depended on three things: the wind speed, the shape of the body of water involved, and the length of time over which the wind had been blowing. They then turned their attention to surf, what happened to waves as they approached the shore. This information was made available to the Navy both in written form and in lectures on military oceanography given at Scripps Institution of Oceanography, where Sverdrup was director.[14]

The practical relevance of all these lines of research for the military is readily apparent. The significance of it for the war at sea was described by NDRC's official historian this way:

> Division 6 [which studied the ocean] was the only one in NDRC to be given objectives as broad as a whole field of warfare. Its activities, as has been seen, covered the whole field of subsurface war, including methods of selection, training, installation, maintenance, and operational analysis, as well as the development of new weapons and devices. Throughout the whole field the NDRC scientists acquitted themselves superbly. Nowhere was collaboration with the armed services closer or more fruitful. The triumphs were a joint product of the NDRC and the Navy, jointly conceived, produced and applied. They went to the heart of two of the greatest problems of the war: the defeat of the U-boat, without which victory could not have been achieved at all; and the strangulation of Japan by sea power exerted beneath the surface, without which it could only have been won much later and at much greater cost of life.[15]

Just as the military outcome of the war was transformed by the collaboration of scientists and the military, so were the structure of scientific research and many of the intellectual priorities of researchers. Many scientists had developed research skills and experience during the war that changed their careers. This is particularly true in oceanography because the field did not really have much of an institutional existence until the Navy gave scientists reasons to invent it. Engineers, physicists, and biologists found themselves functioning as marine scientists during the war, and used funds after the war to continue this direction in their careers. Institutions of oceanography were expanded and became major research facilities in which these scientists could work. The research itself became less dominated by immediately pressing strategic problems, and the authority given the scientists to use research to serve science as well as the state was increased, resulting in the soft-money support system for ocean science.

DEMOBILIZING RESEARCH

What was the role of scientists at this point? On the one hand, they were made a kind of pampered ward of the state, hoarded as a national asset and ready to maintain or favorably shift the balance of international power. Of course, on the other hand, they were state dependents and only supported when they were willing and able to follow research programs deemed valuable by government officials. This situation seemed beneficial enough for most scientists. More funds were put into laboratories and research projects than ever before, and that seemed to enhance the power and autonomy of science. Fewer strings were also attached to a larger portion of funds than during the war. But scientists were not given any autonomous political voice to equal their strategic importance, and they were also not given as much freedom of choice in research as it first appeared. Researchers who had worked in the military or in military-run labs now found themselves in private institutions on government contracts (if they were lucky). This gave them substantially more freedom to design research projects to enhance science and their own careers, and since they were used to working for the government, they accepted as inevitable that the state would use the power of the purse to direct the general movements of science. At least they could pursue basic research in the areas in which there was strong government or scientific interest. They could function as a *reserve* labor force, cultivating potentially useful skills, without being asked to provide immediately useful work for the state.[16]

This change in funding practices came with a change in the social organization of government funding for science. The wartime agencies that funded scientific research were dismantled, and even military funding for science was reorganized as part of a fundamental reorganization of the military bureaucracy itself.

With the surrender of Japan the NDRC (National Defense Research Committee) was getting out of business. It was up to the Bureau of Ships to take over the things that we wanted to keep on going. That was one of my jobs, to arrange that. X and I wrote a letter to the chief of the Bureau of Ships, not *to* the chief, *for* the chief of the Bureau of Ships to sign setting up something called the Marine Physical Laboratory at Point Loma, which would be led by Carl Eckart, the great physicist from the University of Chicago. This was an interesting experience because the Navy had never supported research before—real research, except at the Naval Research Laboratory. Moreover, they didn't like to make long-term commitments. Carl Eckart insisted that they couldn't give us a year-to-year contract. I remember Admiral Cochrane couldn't make up his mind about this for quite a long time. [Interview with a distinguished professor and designer of the postwar ocean research system]

The movement away from wartime mobilization of scientists toward the peacetime cultivation of scientific skills was important for the establishment of an elite reserve labor force in the period. Many military men were reluctant to give up the degree of control they had had over scientists during the war. To the extent that it could, the government reestablished control. But the military was not in a position to keep a large group of scientists mobilized indefinitely, and even partially demobilizing them meant losing some control over their work.[17]

The result was that many laboratories and scientists that had worked in the war effort now began to apply for federal funds to do more basic research related to their wartime work. They continued to cultivate their ties with the military, but on a new basis that made the researchers more collaborators with or consultants to the state than government employees.

During World War II University of California was involved, operated the . . . University of California Division of War Research. So it was basically an acoustics laboratory. And at the end of the war, the university decided they didn't want to run it anymore. And it was sort of run like Los Alamos is, but the University voted not to, elected not to continue operating the laboratory. And so the basic operation there was transferred or combined with what was called the U.S. Navy Radio and Sound Laboratory, which was located here in San Diego back there on Point Loma, and became the U.S. Navy Electronics Laboratory. And . . . most of the people that were employed in the UCDWR transferred over to the Navy. However, the bureau of ships at that time decided they would still like to maintain a connection with the University and so the MPL was set up eventually as an organized research unit. . . . We had sort of block funding support out of "Bu" [Bureau of] ships. And ONR provided some funding. We had quite a few projects that were supplemented by ONR funds and then "Bu" ships decided that they weren't gonna be in the business of it and ONR took it over. Then we had "Bu" ships funding and ONR core support. . . . The big supporter in underwater acoustics has been the Navy. They use sonar and MPL here has worked, been primarily involved in two areas: [acoustic] instrumentation on one side, and looking at the environment, the acoustic environment, on the other. . . . That's one of the reasons why we have space out at what's now the Naval Systems Center, which used to be NEL. And initially, we actually were a division of [the university] and also a department of NEL. We received . . . a large amount of our actual materials support, and shop support and all, out of NEL as just a department or a division of NEL. And then, gradually, why, we moved away from that. We had space in the building there. And we have maintained always good liaison with many of the people within the laboratory. We still continue cooperative projects and we talk with them. . . . So it's been a historical relationship with them. [Interview with a senior scientist at a major research institution]

This story describes a new relationship between MPL and the Navy after the end of the war that was common for laboratories in the period: decreased military control of research but continued military funding and scientific collaboration/consultation. The result was the kind of work environment for scientists in which they continue to advise the state about how to use natural forces and resources to advance state power.

Other laboratories that had thrived with wartime budgets also took advantage of postwar funding opportunities. Woods Hole Oceanographic Institution, for example, was encouraged to keep its alliance with the military both by the military and because of inflation. Before the war, the institution was open for research only during the summers; the scientists would come there after teaching at universities during the school year. But inflation made it too expensive to run the institution for only three months. The laboratory could not maintain the real estate without the year-round presence of the scientists (as was possible during the war), and the institution's budget rose enormously. Government funding finally solved the problem, with the government giving the researchers at the institution enough money to enable them to give up their university positions.[18]

Research after the war may have been less strategically directed than wartime research, but the funding system supporting it was brought to life in large part by Cold War anxieties about military preparedness. Many oceanographers who entered into the soft-money funding system were first recruited to study the atomic tests in the Pacific.[19] Operation Crossroads, the atomic test at Bikini Island, was the most dramatic example of military funding of ocean research in the service of the Cold War. It was a training ground for many young ocean scientists, who developed follow-up studies of wave motion and radioactivity in the oceans, or who continued to work as technological innovators, finding new ways to measure the properties of the ocean, including the use of acoustics. A whole generation of ocean scientists was composed of researchers, most of whom had known each other at Bikini. There they immediately found themselves with difficult intellectual and political puzzles to solve.

There must have been a thousand people in our oceanography group. . . . Oceanographically, we thought that there were really three problems, as I remember it. One was to measure the waves that were created by both the air drop bomb and the underwater bomb. There were lots of quite different ideas, diverse ideas about how big those waves would be. The second problem was to follow the radioactivity, both inside and outside the atoll, the diffusion of that radioactivity. The third problem was the effect on the organisms and the atoll itself, on the coral reef itself, of the explosions.

. . . We did a lot of work on all three of those problems. To give you some idea of the complexity of it, in measuring the waves we set up three photographic towers, 100-foot towers, with automatic cameras on them. The pictures that you've seen, particularly of the underwater test, that huge . . . base surge— those were all taken by those cameras of ours.

. . . If you look carefully at those photographs, you'll see a base surge, you see a kind of mound of water moving out—what looks like water but is really a spray—moving out. But we never noticed that [before]. So when the underwater explosion went off, there was this huge pillar of water, really just mostly spray, that went up into the air. You couldn't really tell what it was, but it looked like a great, huge column of water shooting up into the air. Then it settled down again and went out in this base surge.

. . . The base surge, as I remember it, was several hundred feet high at first and moving very fast. We thought it was a huge wave. In other words, our estimates of the waves had been completely haywire. But it moved more and more slowly and after a while it just stopped. And then after a while it lifted from the water altogether. All it was was spray, a lot of spray; but of course it rained out, and as it rained it became lighter. Finally it got so light that it just lifted from the surface altogether. This was a real phenomenon, this spray. It probably had a lot to do with covering a lot of the ships with radioactive material. Because it just came down on it, yes. . . . It spread out over an area of about a mile, as I remember it. So a lot of the ships were covered with radioactive water.[20]

What was the role of the scientists at Bikini? The historian Susan Schlee argues that the ocean researchers entered into a "workfare"-type program, finding a position on the government payroll to keep them from abandoning ocean research after their demobilization. Roger Revelle, who was in charge of the scientific expedition at Bikini, has argued that Schlee's analysis does not do credit to the seriousness of the science done at that time. But if we accept that the research *was* serious and also admit that the program *was* a way to keep scientists with skills of interest to the Navy in business, then we can see the scientific mission at Bikini as an early government attempt at establishing an elite reserve of scientists. From this point of view, the government hired the scientists to do research at Bikini less because they had need of scientific information from the research than because they wanted to have their own scientists knowledgeable about the effects of nuclear blasts. This cadre could then help them in setting up military strategies for a nuclear age.

The research done at Bikini is important because it demonstrates how scientists began to use their research to serve simultaneously the interests of the government and the goals of science. The projects that the scientists conducted are distinguishable from earlier research done by oceanographers during the war because the avowed purpose of the research, to understand the effects of nuclear blasts on the lagoon, was both broad and more broadly interpreted by the military men managing it than most applied research during the war. We can get a sense of how novel this policy seemed to researchers at the time from a notecard Walter Munk wrote during the mission (presumably to give a talk about it). There he described the "General Directive" of the research:

Study environment as it may effect bomb experim[ent]
Study effect of bomb on environment
To do so, must know what physical [and] biological [environments are] like
To be reasonably certain [we need to make sure patterns are] not seasonal,
    control Rongelap [lagoon]
Credit to Navy for [the] *liberal interpretation* [of the] directive
Within 2 months after we arrived, [we] knew more about Bik[ini] than any
    comparative area in USA
Reason: *Combined Attack*[21]
(The text italicized here was typed in red on the notecard)

In three different ways, Munk suggests that the rules of research have somehow changed with the Bikini expedition, allowing scientists to plan their research to serve scientific theory as well as the military. He is allowed to develop a project for studying circulation in the Bikini lagoon that provides more than the military needs for understanding the movement of radioactive water from the lagoon; he "credits" the Navy for giving scientists on the project enough freedom to interpret their "directive" freely; and he lauds the virtues of this collaborative work in yielding better science.[22]

Some of the scientific riches Munk gained from the research are apparent in a paper he wrote on the expedition, entitled "The Circulation System of Bikini and Rongelap Lagoons." In it, he presents his new discovery of a secondary circulatory system in the lagoons, consisting of a set of two "counterrotating compartments which move in a clockwise sense in the southern portions and in a counterclockwise sense in the northern portions."[23] Reading this paper, there is very little evidence that the research resulted from a military mission. Of course, Munk is not oblivious to this fact. In another paper from the same period, he says, "the slowness with which the lagoon is flushed means that only small amounts of contaminated water enters the ocean at any time—direct outside contamination by rain storms will be a more important factor, and the efforts in contamination should be concentrated towards this end."[24]

These two documents are fascinating to study side by side. They present Munk's two approaches to the study of water circulation, the scientific and the military one, and reflect the dual meanings of the project to this man. Moreover, they show how a scientist could use government funds from an applied program to look for significant scientific findings, while also developing exactly the kind of expertise to advise the government about radiation.

The idea that science and the government share common interests in the results of basic research is both supported and belied by this example. Both Munk and his funders had interests in his developing some expertise about the circulation of the lagoons, but the military did not care if he discovered a new circulation system in the lagoon; it only cared about what the circulation implied about the flow of contamination into the ocean from the lagoon. There was a partial separation of the interests of scientists and the state in the information derived from the research, but a coincidence of interest in Munk's expertise. This mix of mutuality and difference in relation to research was at the heart of the soft-money funding system and new to the postwar era. What did it mean in this period to say that scientists had greater control over their research projects or that they gained greater autonomy? It meant that scientists were encouraged to design research projects that would simultaneously provide them with results that could have scientific significance and give them the skills and expertise to be useful counselors to the state.

To make the contrast between this situation and conditions of wartime research, one can look at, for example, the University of California Division of War Research papers saved by Russell Raitt. First of all, Raitt has few papers in his files describing basic science. Most describe applied projects in progress at the UCDWR to further the war effort. What is perhaps more telling is that the writing on basic science bears no relationship to the research being done at the lab for the Navy. Why is this? To a large extent it is because military research was classified. But it is also because during the war, many scientists did not try to make their work serve science; they wanted to end the war with an Allied victory just as much as the military did. Scientists also often did research that was not rigorous enough (methodologically) or was too practical in its purposes to support scientific generalization.

Roger Revelle's papers from the same period show a similar lack of attention to basic science and little connection between war research and advances in science. Not only did Revelle's rate of publication diminish during and particularly just after the war, but all but one of the few articles he published in the period bore no direct relationship to his military activities. While by 1943 the Navy was sending him around the country to oversee the deployment of new acoustic technologies (mostly bathythermographs) for submarines, he was writing about evidence of seawater in ground water, current measurements in the North Pacific and off the California coast, and marine bottom samples.[25]

After the war, scientists increasingly began to develop identities and a sense of purpose that were less closely tied to military interests. Sometimes this was expressed in political mobilization of the sort visible in the Scientists' Movement, wherein scientists tried to affect military uses of research. More often, it led scientists to think more about what scientific riches they could reap from the pure or applied research that they did for the state. The liberal interpretation of the mission at Bikini noted by Munk was an opening for this separation, a refreshing chance for scientists who were used to working for the military in the wartime atmosphere to think more about ideas and findings of scientific significance.[26]

The novel legitimacy of doing basic science this way was what Revelle likes to emphasize about the research done at Bikini. Scientists were allowed, even encouraged, to make the most of this and subsequent expeditions by collecting some theoretically important data. Revelle fought to make this the case and used the situation to his own distinct scientific advantage, studying atolls to confirm Darwin's theory about their formation.[27]

The freedom given scientists in this system may have been pronounced in contrast to what they had experienced during the war, but it was far from complete. Scientists could do basic research on the stipulation that they chose topics of interest to the military or other agencies of the government. Basic research was encouraged, but only as a means of strengthening the state, not as a means to promote science. The applied research at Bikini, for example, helped to document the nuclear strength of the United States (or at least an illusion of strength) and to create the reality of nuclear armament in the nation that the military sought. At the same time, it taught scientists to become dependents of the state. They became clever at designing research agendas that would yield solid scientific gains for their labs and their disciplines while also touching on issues vital to the state. In other words, they learned how to sell themselves as consultants and combine their consultation with research. They learned to function as trained manpower for the state, while the state learned how to cultivate and mobilize that labor pool.

The establishment of more long-term funding for ocean research by the Navy, through the Office of Naval Research, continued this pattern, making oceanographic research (and a skilled labor force of marine scientists) more stable and legitimate. Scientists were encouraged to tackle basic research problems that were significant to scientific theory; in this way, the Navy was able to keep its favored scientific consultants in best intellectual form in case the Cold War heated up.[28]

Soft-Money Funding of Basic Research

The Office of Naval Research, established in 1946, became the first agency to be successful in providing long-term soft-money funding for basic scientific research.[29] As we saw earlier, the United States was demobilizing in this period, but the military did not want all the scientists to be let go; there was a remilitarization rather than a demilitarization in the period that made the services of scientists seem necessary. At the same time, the military could not use an endless series of Pacific tests or other applied projects for this purpose. So permanent funding had to be established.[30]

The Navy in particular felt the need to keep scientific consultants on hand during this period. The Navy was in a very vulnerable position itself at the end of the war; it seemed that the future of the military in the nuclear age lay with the Air Force. Submarines still remained a potent conventional weapon and assured some future for the Navy, but the success of submarines depended on the superiority of American ocean science. Submarines were lethal when they could hide from surface ships and when enemy submarines were vulnerable to exposure. This situation would be possible if the Navy knew more about acoustics and the ocean than military men in other countries. As a result, much Navy money was spent to keep a body of scientists involved with the most advanced research (using both applied money for acoustic research and ONR funds for more basic information about the ocean). Eventually, the success of this strategy made submarines an increasingly attractive place for locating nuclear missiles, insuring the future of the Navy and providing a permanent rationale for funding ocean research.[31]

Given this context, it makes some sense that the Navy would have initiated and later pushed for ongoing government fundings of ocean research. In this scenario, scientists were kept on the government payroll not so much for what they knew but for their ability to keep the U.S.

Navy more informed than anyone else about the ocean and about gathering information in the oceans.

Obviously, the military also continued to support much applied research done in national or military laboratories. In fact, most of the research supported with military money was supported outside ONR and remained applied work. But that does not detract from the precedent-setting character of ONR. It was the first source of regular research money for pursuing "basic" research.

One reason why the ONR model of scientific support was tried and eventually copied in other agencies is that, while it was run by the Navy for the Navy, ONR was not simply invented by Navy personnel. It bore strong resemblance to a program proposed by Vannevar Bush to President Roosevelt that suggested how the government might profit in the postwar era from scientific research. Bush had been the head of the OSRD during the war, realized the importance of science to the war effort, and was determined to dismantle the OSRD without letting the national science effort started in the war period die out. That was exactly what had happened at the end of the First World War, and many people did not want it to happen again. The danger of abandoning science at the end of the war seemed all the greater after World War II because the power of science and the contributions of scientists to winning the war were much more visible. The use of the A-bomb had even made scientists celebrities and their contribution to war a matter of public interest. But even before, many people in the military attributed great importance to science.[32]

Bush's proposal, published as *Science—The Endless Frontier*, advocated support of basic research in independent laboratories. Bush suggested the government should fund science to bolster medical care, economic growth, and military advantages in the United States, and also to reduce the strategic vulnerability inherent in depending on European science. Clearly, he saw scientists as useful advisors whose skills should be kept honed by the government.[33]

He also advocated a decentralized research system. This would put funding agencies in touch with diverse elements in the scientific community, and therefore keep the government from being misadvised by a small clique of politically skilled but scientifically limited researchers. More practically, sending funds to a variety of states would ensure that many congressmen would support a substantial budget for research. There were also some less political payoffs from the system. If a variety of laboratories could be supported to pursue different research directions, the government did not have to know in advance which would pan out; it could (and did) make decisions about which directions were *unlikely* to be worthwhile and thus not worth funding, but it could back more than one "horse" in any given scientific field. Given the size of the United States and its economy, this kind of strategy was feasible (which it was not really in Europe); on the ideological level, this approach to funding was appealing because it seemed more democratic.[34]

Alan Waterman, the head of NSF when Bush's proposal was reprinted in 1960, underscored the continuing importance of the Bush view of scientific funding as a rationale for state funding. In his introduction to the book, he said NSF was reproducing the text "not as an historical document, but as a classical expression of desirable relationships between government and science in the United States."[35] Here is a little of what Bush said:

> Basic research is performed without thought of practical ends. It results in general knowledge and an understanding of nature and its laws. This general knowledge provides the means of answering a large number of important practical problems, though it may not give a complete specific answer to any one of them. The function of applied research is to provide such answers. The scientist doing basic research may not be at all interested in the practical applications of his work, yet the further progress of industrial development would eventually stagnate if basic scientific research were long neglected. . . . Basic research leads to new knowledge. It provides scientific capital. It creates the fund from which the practical applications of knowledge must be drawn. New products and new processes do not appear full-grown. They are founded on new principles and new conceptions, which in turn are painstakingly developed by research in the purest realms of science. Today it is truer than ever that basic research is the pacemaker of technological progress. In the nineteenth century, Yankee mechanical ingenuity, building largely upon the basic discoveries of European scientists, could greatly advance the technical arts. Now the situation is different. *A nation which depends upon others for its new basic scientific knowledge will be slow in its industrial progress and weak in its competitive position in world trade, regardless of its mechanical skill.* [Italics in original text][36]

The Bush document simultaneously idealizes science and its practice, using a language that treats science and scientists as completely detached from practical concerns,[37] and upholds the state's interest in supporting science because of its practical usefulness. The ideal image of science suits the culture of scientists, flattering them and underscoring their socially high status, and the pragmatic approach to the value of science makes it clear that the government is not to give money to science from a sense of charity or a feeling of obligation developed during wartime.

Bush's arguments about the importance of science to military preparedness, like his ideas about science and economic growth, share the same duality. He draws a distinction between applied work done within the military and the basic work that independent scientists must do, but on both levels he emphasizes the scientist's contribution to a state's competitiveness. He assumes a natural coincidence of interest between scientists and the state, just a difference of temperament and skill between the two.

> Some research on military problems should be conducted, in time of peace as well as in war, by civilians independently of the military establishment. It is the primary responsibility of the Army and Navy to train the men, make available the weapons, and employ the strategy that will bring victory in combat. The Armed Services cannot be expected to be experts in all of the complicated fields which make it possible for a great nation to fight successfully in total war. There are certain kinds of research—such as research on the improvement of existing weapons—which can best be done within the military establishment. However, the job of long-range research involving the application of the newest scientific discoveries to military needs should be the responsibility of those civilian scientists in the universities and in industry who are best trained to discharge it thoroughly and successfully. It is essential that both kinds of research go forward and that there be the closest liaison between the two groups.[38]

Civilian science is depicted here as a world of experts that the military cannot easily create for itself. It is a resource to be used to promote the military strength of "a great nation." The prose still emphasizes scientists making "discoveries" (i.e., being productive), but it also suggests that they may consult with admirals or generals too.[39]

Bush urges that civilian science be funded by a government agency adhering to the following principles: (1) funding would be long-term to assure the possibility of long-term research; (2) the agency would be in the hands of citizens (presumably as opposed to the military and perhaps politicians); (3) the agency would promote science through grants and contracts to organizations outside of government (primarily universities and explicitly not government laboratories); (4) funds made available to researchers in universities and colleges would not be used to interfere with the curriculum or larger research agendas of those institutions; and (5) the agency would retain the power to decide how to allocate funds and audit their uses.[40]

The first four points all make the system appealing to scientists by giving them some autonomy from the government in general (particularly the military), but the fourth section declares the fundamental right of the government to control the funding system. The elite status of scientists is again acknowledged in the partial autonomy from direct government supervision of research, but the right of the government to control directions of research and the manpower pool engaged in research through controlling the purse strings is maintained.[41]

The model of funding in the Bush report shaped policy at ONR, and, through ONR, affected the other funding for basic research developed later by the government. The rationale for funding remained, on paper, the applicability of basic research and the productivity of scientists, treating scientists as a precious elite with clear social value. Yet the fundamental mix of autonomy and dependency was also written into the system, serving the manpower interests of the state in scientific expertise. Thus the role of scientists as an elite reserve labor force was forged through the Bush proposal and ONR's partial implementation of the Bush system.

This is not to say that academic scientists *only* received funding from ONR and *only* conducted basic research in the period. ONR is emphasized in the histories of the period because of its novelty, but the agency was not the only source of support for science in the government or even the military. There were many who had military contracts to do applied research for other parts of the Navy or the military. Some scientists worked with a combination of funds for both "pure" and applied research. This was possible because the line between the two was less clear than convention would have it. For students of acoustics, for example, the line between pure and applied research has always been vague, a mere change in point of view on the same data. The same scientists in the same laboratories might be simultaneously engaged in research for a weapon system and interested in accounting in theoretical ways for anomalies in undersea acoustics.[42]

The movement orchestrated by Bush from military control of science at the end of the war not only affected the funding opportunities of researchers, but also precipitated changes in the *bases* for allocating funds. The criteria for identifying quality in research changed as the demands on researchers also changed. As the manpower needs of the government started to shift, the mechanisms for shaping the labor pool were (not self-consciously but nonetheless) revised. During the war, for example, research was tied directly to particular military problems. Admirals wanted to know how waves would break on a *specific* beach on a *specific* day so that an amphibious assault could be planned. These specific projects were fleeting and were replaced regularly by new projects as military actions were completed and new ones contemplated. This gave the military interests in locating and using scientists who were broad enough in their education and reasoning abilities to address a variety of different questions as they arose. As a result, during the war, the military tried to cultivate a group of people who were good problem solvers, who knew how to generalize from theory and knew how to apply theory to practice, and they tried to support the same group after the war.

Not surprisingly, once ONR was set up, the Navy continued to rely on some of these wartime values in deciding *how* to support basic research. In the beginning, ONR funded individuals and groups of problem solvers rather than research projects, giving them block funds rather than money tied to particular activities.[43]

One of the things [different] in that era, [was that] the interaction between laboratories and their sponsors was somewhat more cooperative and individualistic in the sense that [for example] Marine Physical Laboratory was funded kind of on a block basis [for] what the total staff produced, rather than [for] individual projects, each of which had to be separately funded. [Interview with a senior scientist in a major research institution]

The Atomic Energy Commission (AEC) also engaged scientists to do research, starting with the pool of ocean researchers who had been involved with the early atomic tests, and continuing with other projects.[44] It is difficult to find much information about this, since most of the research was classified, but informants again emphasize that persons more than projects were the recipients of funding during the period before the establishment of the National Science Foundation. These researchers were supported more for what some describe as their abilities and others describe as their reputations than for the quality of their proposed research projects.

This funding by reputation rather than project began to break down in the 1960s. Fears about technological backwardness after Sputnik created a momentary increase in the money made available to oceanographers, creating a heyday for oceanography in the early 1960s (which had its intellectual expression in the development of plate tectonic theory), but after this period, cutbacks and increased oversight of research began to create the more bureaucratized and commercial vision of research funding that dominates today.[45]

The time when oceanography expanded rapidly was beginning in about 1955 and it ran comfortably until 1967, and then it came to a sudden halt because at that point all of a sudden the first, ah, Nixon administration put an absolute ceiling on all funding. And there was no more, and it was a low ceiling, less than we'd had before so people had not only to cut back, not only not to grow, they had to cut back.

. . . The ONR thing began to change. . . . The Mansfield Amendment said that it had to be of some relevance to the Navy. That began the cut back and then the Navy had to begin to set up targets. You couldn't just go and say yeah, I'm a good guy. . . . ONR was always interesting because they supported people largely on their reputation. ONR near really cared too much about publication and things like that. They had people; they were very enlightened. They didn't open the doors to everybody. They had a bunch of customers whom they regarded as interesting people with good ideas who were hard working and delivered. [You] didn't have to write elaborate proposals, and once you were in, they'd protect you and you would also know that if they finally decided to phase you out, they would give you a chance so that you wouldn't be closed down in five minutes flat like you do if your NSF grant doesn't go through. And so [you] could afford to build something on it and know you wouldn't have to fire anybody in June because the grant didn't come. And they were very excellent to work with. But as the . . . Congress insisted that it had to be close, [to have] demonstrable relevance to the Navy mission, it became much more difficult and the group shrank, and of course they too got a ceiling on their funds like everybody else. And that turned the thing to NSF. [Interview with a senior scientist at a major university]

The turn to NSF was not complete, but the change was still dramatic. In 1958, of $8.3 million of federal money for university ocean research, $1.6 million was from NSF and $4.5 million was from ONR. In contrast, by 1968, of $41.9 million for university ocean research, $13.9 million came from NSF and $20.9 million came from ONR. Then in the 1970s, NSF funding grew to around $60 million, giving it the more prominent position in shaping U.S. ocean research.[46] (Inflation during that period makes the budget increases seem more significant than they were. Oceanography was well served by some programs at NSF, particularly the money set aside for the International Decade of Ocean Exploration, but the funding picture was in fact less favorable for many researchers than earlier because of the cuts in ONR.)

NSF was charged both with defining and supporting quality research in the sciences (not understanding the ocean as ONR did), and thus it began to change what many oceanographers had to do to secure funds for their laboratories. NSF was much more interested than ONR in the proposals for projects than in the reliability of those seeking money because agency personnel were compelled to consider the *scientific* benefit of the research. Applicants who were well-known scientists in their fields and had a history of successful research were still at an advantage in the search for funds, since they could help define in other contexts what good science was. But now increasingly, scientists had to *demonstrate* their analytic skills in planning and writing extensive proposals for well-thought-out projects. They had to *demonstrate* their mastery of relevant literature and techniques of research as well as argue their way around current controversies in the field. The purpose of this "project orientation" was to reserve support for the *demonstrably* best scientific thinking of the time. At different times, particular areas of research have been favored by NSF, and researchers writing poorer proposals in that area have been able to secure funds. Other anomalies in the system have also arisen, thanks to cronyism or fashions in science, but the commitment to quality control in these agencies has been basically sustained.[47]

The research projects suported by NSF and other agencies that appear to have no apparent practical value still have had an important function for the government as systems of quality control, defining what constitutes good work and who are the scientists doing it. The government needs to know what work has or does not have scientific legitimacy, since agencies wanting to allocate funds for a project need to know how to identify the better researchers for the job.

Agencies like the Office of Naval Research and Department of Energy (DOE, or before that, AEC) have not required this kind of extensive proposal writing. But being charged with demonstrating the explicitly practical usefulness of research, they too have developed a higher degree of oversight over the work of researchers. These agencies have been required to justify their funding patterns more rigorously, so they have passed this rigor on to the researchers. Thus contemporary grants provide less freedom for their principal investigators than before. There is increased accountability both in terms of how money is spent and what scientific or practical benefits accrue from it. Ironically, this loss of freedom has resulted from the shift away from military control of funding (with the establishment of NSF and NIH), but it also resulted from the restriction of funds during the late 1960s and 1970s.[48]

The constricted budgets of the sixties and seventies also affected the skills of scientists in another way; it led to the cultivation of political skills to gain competitive advantage in a world of shrinking resources. Research groups were increasingly pitted against one another for resources, prospering only at another's expense, and flourishing only when scientists were able to act as shrewd politicians and keep the right connections in government.

> When I went to [New Lab], it badly wanted to grow but in a steady universe; it had to go and take it away from somewhere else. And they took it away from the [Wet Lab] which had been supported for many years but was burdened with all sorts of things that made it very ineffective. We brought there a very good, eager, hard-working, heavy-publishing young crew who made a lot of use of students who wrote lots of papers. . . . We got a lot of stuff published by having the students publish it. That doesn't occur to superstars who would like to publish it themselves. And as a result when I came to [New Lab] I think the budget was less than a million dollars, and when I left it was eleven million. And that was before inflation set in, and most of that money came out of the

[Wet Lab], who lost their ONR contract and maybe all of that went to [New Lab]. We really were very successful. But it was done by having your ear to the ground and listening very carefully, sometimes leading to embarrassing situations. At one point . . . they had closed down a NOAA [National Oceanic and Atmospheric Administration] facility, which had a lot of geophysical equipment, and I maintained a wonderful pipeline system to Washington, more or less by going there exactly two days a week for a commute. And I got a phone call and somebody said, "The decision has just been taken to close down the geophysics lab at NOAA. Equipment is surplus. If you can use it, contact the director and we will authorize transfer." And so I contacted the director and he hadn't even heard of it by then. [Interview with a senior scientist at a major university]

These shifts in funding policies could be taken as evidence that the elite reserve labor force as it was set up in the 1940s was disbanded in the late 1960s and 1970s. But they can also be evidence that the criteria for recruitment into the elite reserve labor force simply changed. The government still wanted to support the scientists with the best research skills to keep their research expertise as refined as possible. They just began to measure the skills in a more bureaucratic manner, making decisions on the basis of how well proposals were written and on skills in dealing with agency personnel.[49]

Saying that there was a major shift in this period may also be something of an exaggeration. A number of scientists I interviewed contended that ONR still funds people more than projects and that NSF also makes decisions based (at least in part) on the reputation of those applying for funds. Established researchers who review proposals for NSF say that they take the abilities of established scientists into account when they review one of their proposals (although they will be happy to turn them down if the proposals are bad). And young researchers applying to NSF also firmly believe that they are less likely to be funded for the same quality of proposal than a more established researcher. The Cole, Rubin, and Cole study of NSF suggests that reputation and lack of funding history do not affect funding decisions very much, but they still do affect them. Researchers who have gotten funds are more likely to get additional funding than researchers who write an equally good proposal but have not received funding before. Perhaps this slight advantage is enough to make funding by reputation a reality to the scientists involved.[50]

The limits of the "project orientation" also show up in other ways. One scientist I interviewed tried to change the direction of his research, and submitted a very tight proposal to NSF. He was then told by an NSF official that he had no track record in the new research area, so he could not get NSF funds to move in that direction. It did not matter what his proposal looked like. The decision was made based on an evaluation of how the proposal fit his research history. This kind of evidence suggests that elaborate proposals may now be the primary measure, but still are only one of many used to identify those who deserve to receive state funding.[51]

Through all the changes in the funding system since the 1940s, the basic Bush strategy has been maintained. Government officials have continued to try to identify the more worthy scientists and support their basic research. Each change in policy by government agencies has led to some shifts in the type and amount of research funded. Hence labs have waxed and waned along with different types of projects. But the role of government as a source of funds for research has not been repudiated, and the identification of national progress with basic research, although sometimes under fire, has remained a powerful argument and the one primarily justifying continued research support.

Through grants and contracts for both applied and basic research, then, the government has helped to create a body of scientists who are able to devote their time to research. Just as the military had created a pool of wartime talent, postwar agencies have maintained a peacetime group prepared to inform the country about the nature of the ocean and the use of its natural forces and resources to empower (or at least help maintain the power of) the state. Not all of them have been asked explicitly to provide government service, but their work has informed the advisors who have transferred their expertise to the state. They have helped, in this fashion, to build the nuclear Navy in the United States, the nuclear power industry, offshore oil drilling, and other military and economic systems for exploiting both knowledge about and mysteries of the ocean.

CHAPTER FOUR

Managing the Scientific Labor Force

How GOVERNMENT agencies spend their research money on scientific projects shapes the character of the U.S. scientific labor force. For both inside and outside funding agencies, it may seem that the allocation of research funds is more random than people would like and so variable across agencies that it never could constitute a coherent state policy. But there are some fundamental assumptions about how scientific research should be conducted and about the government's relationship to science that give rise to clear patterns in the funding decisions made by different government agencies. That is why, although there is little centralized state planning of scientific research in the United States, there is still some underlying science policy, and through the process of dispensing monies, the U.S. government manages its scientific labor force.[1]

Government funders can manage the scientific labor force only because they take advantage of surveillance systems developed by scientists and agency personnel to regulate their relationships. We can think of the funding system as a communication system, in which government agencies display their funding priorities and scientists display their research capacities and plans. If you look just at the proposals and research reports sent to agencies, it looks as though the center of the communication system lies in scientists writing documents for the government about their proposed and actual projects, but that is not the whole story. Even more information is gathered informally by scientists and agency personnel. Scientists use their connections to figure out how to make their proposals as fundable as possible. Agency personnel in turn learn about scientists' research skills, reputations, and scientific expertise to help them choose reviewers for projects or to determine how to spend their limited budgets wisely.

At NSF, the center of this surveillance system is in the "rotators," scientists who work (primarily as program officers) for (usually) two years at the agency, and then return to their laboratories and universities. At NSF, they are expected to use their knowledge of their disciplines and the people who work in them to keep the agency on top of the latest developments in science. Rotators are hired to help the agency figure out how to evaluate research proposals and to champion particular (and sometimes new to NSF) directions in research. For the latter purpose, they dispense "insider" knowledge of NSF to their colleagues by letting them know what kinds of work they are advocating in the agency. In this way, they have a chance of generating "proposal pressure" on the agency to increase kinds of research that rotators favor. Proposal pressure means a rapid growth in proposals that strains NSF's ability to fund in an area of research. NSF people often read dramatic variations in proposal-writing activity (which often occurs spontaneously) as indicators of changes in science to which they should respond. Program officers, recognizing the legitimacy of these measures within the agency have been encouraged to stimulate a proliferation of proposals in the areas they want to fund. In this way and others, rotators convey information both to scientists about NSF and vice versa, thereby shaping NSF funding patterns. The fact that many of them are able to suggest their replacements at the agency makes it easier for them to keep the communication networks between agencies and scientists stable over time, and makes it all the more central to shaping the characteristics of the scientific labor force supported with NSF money.

Information dispersed through present and former agency personnel about funding predilections is used (as one might expect) primarily by individual scientists or groups of scientists to try to maximize their funding chances, but it is also used by institutions to keep their research staff as active and scientifically distinguished as possible. Administrators at universities and large private research institutions try to cultivate ties to Washington, not just at the level of program officers, but at the level of science policymaking. They try to figure out how to make large grant proposals seem particularly attractive to the government, and how to use their own funds to sweeten the pot. They also try to lobby for programs they think are essential to their research groups, and keep their networks to the world of science policy as strong as possible so they will not miss any unusual funding opportunity they could use to their advantage.

On the formal level, agency personnel also have enormous access to information about people in the scientific community through the myriad applications and research reports they collect from scientists, which constitute written records of their institutional locations, skills, research histories, and the like. That formal level of surveillance is rarely, if ever, used to locate scientists for agency reviews or national policymaking. But even nonrotating personnel at funding agencies do have at their fingertips all kinds of informal indicators of the research areas and research skills to be found in the scientific community.[2] They get some information from listening to the succession of rotators who come through their ranks. Regular agency staff also become acquainted on their own with a large number of scientists. Some make site visits to individual laboratories (to see if they are adequate to support proposed research); others bring scientists to Washington to work on review panels for agencies, or talk to researchers on a more informal basis at scientific meetings or in Washington. Mostly, they use this information to get their jobs done, but sometimes they can use it to help mobilize members of the reserve labor force. When someone within an administration wants to know whom to ask about letting kids with AIDS go to public school, someone at the funding agencies should be able to give him or her a name. And when an admiral wants the Navy to apply new developments in robotics to its program of deep ocean surveillance, someone can provide a list of people to call.[3]

There are two aspects to the surveillance of scientists by agency personnel. One is monitoring the distribution of expertise in the scientific labor pool (a kind of skill topography of the scientific community). The other is monitoring the quality of work done by scientific researchers on their agency's funds. From the point of view of agency personnel, they do not function as a kind of Big Brother, looking over the shoulders of researchers to keep track of what they are doing and to determine if it is worthwhile. They simply want to keep on top of new directions in research, and they want to make their distribution of funds generate the best possible science in the areas they deem worthwhile. This ends up giving them informal, cognitive maps of the scientific community they oversee. They look for who is doing interesting work, and what makes the work interesting. Not every researcher is visible on this map. Some write undistinguished proposals, do undistinguished research, and are never thought of again until they apply for new funds. Since there are so many different agencies, some researchers may never be known well by any government official because they move among a group of agencies in seeking funds and do not make much of a scientific or personal reputation while doing so. Others may be well known at NSF or DOE but not really elsewhere.

Although there is a surveillance system, it is neither formal nor a centralized one. There is no identifiable "state" center for watching scientists and monitoring their research careers. Still, routinized means for overseeing behavior should be called a surveillance system, when by "watching" behavior, it changes it. Since the monitoring of scientists clearly affects the way they seek funds and the way agency personnel make funding decisions, it should indeed be called a surveillance system. Only a limited number of research directions are supported by the state, and scientists tend to flock to these areas. But importantly, the surveillance system also shapes government actions. Decisions about what should be funded may be partially determined by government needs, but it is also strongly affected by impressions of science. The rise and fall of subfields are aggravated by agencies as they "read" trends and then make them happen by the way they channel their money.

According to the Vannevar Bush formula laid out at the end of the war, scientists were to do research outside the direct control of government laboratories, but the government retained the right to control and monitor the use of funds. This way the government could be sure that its money was not wasted. This rather unspecified right provided the state with a legitimate means to use funds to shape the direction of research and monitor the activities of scientists, thereby insuring (at least to some minimal extent) that scientists on government monies would cultivate appropriate skills.[4]

The Bush vision of a centralized system of money for science has not been realized, in spite of the fact that the NSF has been established to support basic research because, when that fledgling organization was established, other agencies did not want to give up their authority and budgets to NSF. They have continued to support their own researchers.[5] Yet there is still some consistency to the funding system. All these organizations, despite differences in charter, philosophy, and funding practices, use most of their pots of money to support a group of experts who can define "state of the art" science and who are capable of assessing the scientific feasibility or advisability of government policies. They provide highly authoritative voices both on the status of scientific knowledge and the distribution of scientific expertise, and also on the natural resources available for enriching the state, on the proper handling of hazards in the natural world (from disease to pollution), and on the use of forces in nature for the pursuit of state power.[6]

PREPARING SCIENTISTS TO BE PROBLEM SOLVERS

What kinds of skills are cultivated and monitored in this system? Some of the important ones are both very obvious and very mundane. All you have to do is look at the array of program areas supported at NSF to see some of them. The government makes sure that at least some minimal group of researchers in the basic sciences and the major subfields within the sciences are supported through NSF. The funding may be small or great, depending on the vitality of the field and its political significance, but the commitment is written into the organization of NSF. Other agencies also have their own lists of types of research that they support, and they too reflect the range of skills that the government is committed to supporting. These institutional measures provide some clues about the skills the government wants scientists to cultivate, but they do not reveal all of them. Some methodological and practical ones cut across a wide range of programs and research subfields. One of these that is almost ubiquitous to oceanography is the skill of doing research at sea.

It might be difficult to go into a laboratory of a gene-splicing specialist and argue that what the technicians were doing there was developing techniques of use to the state. Clearly, enough efforts have been made to use these techniques for medical or agricultural purposes so that one could make this argument, but it would be difficult actually to see this objective in the day-to-day operations of the lab.[7] This is less true when looking at researchers at sea. As one informant made clear to me, just knowing how to do complex studies of the ocean at sea is itself a skill desired by the state.

The reason they fund oceanography is they were very impressed with the few people in the world that knew anything about the oceans during the Second World War—were able to find submarines acoustically. Their attitude is they don't care what I do. I don't do anything with fleet interests. I've never talked to any admiral, or blue-suiters as they call them. I deal with a civilian program manager. And I don't ask them why they fund me, and they don't tell me. But I know what it is. They are always afraid that something is going to come up. *They are going to need smart guys with a lot of experience at sea.* And the way they operate the U.S. Navy, they don't make those guys in blue suits. The Navy is a great towering pyramid of total incompetence. You have no idea. So they keep us on a string so that if push came to shove, whether we liked the Navy or not we would not have very much option if they said the chips are down, baby. [Italics added] [Interview with a senior scientist at a major university]

Having experience at sea may not seem much of a skill, but as the people who do it will tell you, doing open or deep ocean research is no easy feat. Delicate laboratory equipment is usually not designed to withstand the salt air and jostling of shipboard life. And many of the scientists who tend them are also unprepared for the physically difficult environment of a research vessel at sea. In a parallel fashion, just as the instruments designed for research at sea must be specially designed for that purpose, so the scientists who work at sea must also learn a new set of skills to do that work. Some of these skills have to do with using winches and cranes to put instruments into the sea and take them out; others have to do with living in cramped quarters with a large group of other scientists and working in minimally equipped labs on shipboard. No matter how mundane and how unrelated to the intellectual life of science, these skills are still essential for the manpower needs of the Navy and other groups in the government.

Having skills in doing research at sea means having enough experience with problem solving on shipboard that the researcher is able to think seriously about a problem and not be distracted by the foreign sights and sounds of the open or deep ocean. Problems that can arise from using inexperienced researchers to study some phenomenon are illustrated in the descriptions of some early *Alvin* dives to the hydrothermal vents on the deep seafloor. A very large group of biologists and geologists were sent on a cruise together to assess the importance of and do initial research on the vent systems. Unfortunately, the novelty of the experience of diving to the seafloor itself often distracted them from doing serious research. One person described it this way:

> You had had a group of scientists that you respected who had clearly proven themselves and you'd be sitting around the table, planning a mission, as scientists. What are our objectives? What do we want to do? What's the critical observations to make? Where should the samples be made? . . . They performed admirably as scientists to that point and then, who's going to dive? Well, I want to dive, I want to dive, I want to dive, I want to dive, I want to dive. And you'd say but who *should* dive? Screw that, I'm gonna dive, you know. And so what you saw was this humanness that wasn't supposed to be in science, but was. And the warfares that emerged over who was gonna dive was saddening. . . . It seemed wrong. . . . Your first dive, your brain's blown away by diving, you know. . . . You look at the tapes of a geologist, a trained geologist, going, "Look at that yellow fish. My God, wow, look at that, oo, get a load of. . . ." And you're going, "Wait a minute." But that type [of scientist has] gotto go through that phase of bioluminescence, of having eight tons per square inch of pressure, whatever, on your outside sphere to go through the phenomenon of diving. It takes you x number of dives to get that out of your system. But [it does not go away] if you only make one. And then one. And then one. And then one. And then one. And then one. [Interview with a senior scientist at a major research institution]

What scientists need to do if they are to begin thinking seriously on the submarine is get bored enough with diving to attend to the scientific problems at hand. Scientists who regularly do this kind of work or even those who frequently do research on surface vessels complain about the difficulties of the research routines just as they might complain about commuter traffic. To more seasoned researchers, the *only* excitement of being at sea comes from the fresh data they bring back for study.

> I usually spend a month a year [at sea], but when I started in '69, the first year I was at sea six months. I mean, I paid my dues a long time ago. I figured it out once. I've been at sea a little more than two years in the past fifteen. So I don't go to sea very often. I don't particularly like to go to sea. There's some things I like about it but I can forgo those. Because there's more things I dislike about it. . . . But one of the reasons, in fact, that I came here [and started my

career in oceanography] was because I wanted to go to sea. I had this sort of stupidly romantic view of going to sea, you know, isn't that wonderful? You know, go out on a ship and work, wow . . . it's dumb. I mean, it isn't true, you know? Not for me, anyway. And so I sort of had that burned out of me quite a while ago. But there are times when it's real pleasant. I mean, when the weather's really nice. Going to sea is just like going to work. I mean, people ask me still when they find out what I do, oh, you know, like Jacques Cousteau? And probably, or do you dive? You know, cause somehow they view diving as being a great thing. But if you dive, and I've done some diving as part of research, diving is just how you get to the office. That's all. There's nothing romantic about it. It's uncomfortable. It gets cold. It's dangerous. You know. And going to sea is the same way. The big plus about going to sea is that when that net comes up there is no—I don't think there is any—feeling like that. You get that on deck. It's like Pandora's box, every time. You never know what's gonna be in it. [Interview with an experienced researcher in a senior scientist's lab]

From a manpower point of view, the frustrations of research cruises can be highly productive. For a few moments of intellectual challenge, researchers exercise a range of skills potentially useful to the state. They endure hard work, frustration, and ennui just to practice their esoteric crafts in ways that give them suitable skills.

When they go to sea regularly, scientists face many routine, practical problems of doing research in inhospitable environments. The experience motivates some of them to design new research equipment that can make the work more mechanized. A scientist described his inventing new unmanned vehicles for deep ocean research in the following terms:

I had spent, up to that point, a decade of my life in a submarine on the bottom of the ocean, going to sea four months a year, for a decade, forty months at sea. And I'd seen less than one-tenth of one percent of the midocean ridge. I mean, what a way to go. And the midocean ridge is 40,000 miles long, 72,000 kilometers, covers 28 percent of the planet. It's the biggest feature of the planet. Am I really gonna spend the rest of my life on my hands and knees in the dark with a flashlight, crawling around? . . . I had at that time maybe two hundred deep sea dives, and the thrill of getting up at six in the morning and not drinking anything and crawling in a submarine and making a two-hour descent in a cold sphere, starting to freeze, cramped up and legs going to sleep, getting a migraine headache from lack of oxygen, getting a wet shoulder from the hull, to get down on the bottom to stick my head beneath my legs and step on my feet and crack my head and talk for hours after hours into a microphone and become a machinery and a human digitizer. I must admit the first time a saw a black smoker [a hydrothermal vent where the water comes spewing out in a great black cloud of superheated water through a chimney of rock], it blew me out of the seat. The first time I saw a hydrothermal vent I couldn't believe it.

That represented about 5 percent of my bottom time, if that, perhaps less. Then after four hours, to return to the surface, two hours back up, writing out little clichés for little kids that have sent in their stamped envelopes. Eating a lousy sandwich. Walking out with a migraine headache that just made your eyeballs jump out of your head, cold, tired, to sit down then for two days before you dove again and transcribe all your tapes. [Interview with a senior scientist at a major research institution]

This scientist did not find diving all that pleasurable but found himself technologically constrained to dive if he wanted to do research on the seafloor, so he started working on techniques to do deep sea searches from surface vessels. Certainly in developing these innovations, he made himself all the more useful to science and the government. He gave the military a new system for doing deep sea searches, and he gave scientists new tools for their trade. In that way, he is unusual. Many other scientists develop techniques to make work at sea easier, but they make few innovations that they treat seriously. They do develop personal work routines that solve some of the myriad little practical problems that plague their research. One man I interviewed spoke at length about the rim of a winch he improved because the old one gave him problems while he was trying to use a net to sample organisms in the open ocean. Making a more sturdy winch was hardly the point of his cruises, but his work could not be done without it. There are also numerous occasions documented on the *Alvin* tapes in which researchers give one another tips about how to use the camera better or how to use the manipulator arms more effectively. By creating and spreading small tricks that seem particularly worthwhile, scientists working at sea create an innovative folk culture for addressing their routine problems, and on a regular basis upgrade the skill level of the scientific labor force.

On a less visible level, scientists also learn some social skills for living and working well with other researchers at sea that make them more effective in tackling large problems collectively. Research in the deep or open ocean is usually done in groups, since it is hard to justify all the costs of shiptime without having a full ship and crowded agenda of research activities. Much has been made in recent literature on the sociology of science about the managerial activities of all those who do "big science."[8] What we see in the case of ocean research is (at least from a funding point of view) just a special case of this more widespread phenomenon. Someone has to organize the proposal writing and the administration of the project. Still, some of the routine problems of big science are made all the more difficult by conditions at sea. (This kind of difficulty may be matched in research at the North Pole or in equatorial areas, where weather can make the use of research equipment very difficult or at least different, but it is quite unlike the life in most labs.) Life at sea is simply less under control than life in the majority of labs, observatories, or other research sites; nature can exercise a great deal of control over researchers in the ocean, while in contrast, scientists exercise more control in the laboratory. So marine scientists feel even more poignantly than scientists in most other fields the problems of coordinating large numbers of researchers and their complex and delicate equipment, intellects, and egos. More goes wrong so the level of stress and interpersonal tension can be much higher.

[On this particular cruise] there was, in addition, a language barrier. Most of the French didn't speak any English and most of the Americans didn't speak any French. I sort of worked up to the position of informal chief scientist during the actual operations. I made sure the data was all pulled together. . . . It was a very interesting experience. So we went out and dove and it was a real circus because [we had] one [Blue Lab] ship, one French ship, and two submarine motherships and three submarines, and the French had [brought a ship] full of press, and you know that it was a zoo out there in the middle of the Atlantic Ocean, all these ships milling about. Radio conversations day and night. There must have been something like thirty or forty scientists out there. Endless trouble and we're much too far from a base for a first operation so that constantly, when something went wrong, somebody had to go all the way back to port in the Azores. . . . It would take five days for the submarine to get back there and then back out. They couldn't really repair it at sea. It was really a zoo but extremely interesting. I'd hear these wonderful conversations when somebody'd call from the big ship and say, you people still have any inter-office envelopes? We are fully out. And I thought, my God, is this what I went to sea for? To worry about inter-office envelopes? Years ago I had already felt upset when the director of [Big Lab] decided that he needed a weekly report, and so I really got upset and said that's not why I became an oceanographer. To be in the southwestern Pacific and send a weekly report back to the land. This was work. And you know, we spent three, nearly three months out there plagued by trouble and breakdowns because nobody knew. . . . It was all new, and we didn't really know how to work within everybody's means. [Interview with a senior scientist at a major university]

Scientists who run large cruises are placed in charge of a complex group of scientists with their own interests who can frequently have conflicts with one another. The conflicts are aggravated when the number of scientists included in a project is very large and the conditions for research are particularly uncomfortable. But even during smaller cruises, scientists can find themselves faced with one another while working under unbearable pressure, having to stay up night after night trying to analyze specimens that do not last long or making measurements at night that will guide them in experiments or data collection the next day. Conversely, researchers can sometimes wait a long time before they can get specimens or other data to analyze. Sometimes they are trying to locate specific kinds of objects or specimens that are scarce or located in some difficult-to-find area. They simply must wait until something is found. They may go for days with nothing to do and then suddenly find themselves called in the middle of the night to see some specimens or to examine sightings that have come in. Many describe a terrible fatigue that sets in from the unpredictable and often long working hours. It is aggravated by the problems of resting in cramped sleeping quarters. The constant motion of the boat added to the other physical strains contributes to a kind of daze that they say sets in after a week or so of intense work at sea. They start to feel that they cannot really think clearly anymore, and they worry that they may lose their good judgment and miss some analytical opportunity as a result. Solving problems under these conditions is indeed quite a skill.

Researchers who work on the submersibles also develop a range of social-psychological means for normalizing the process of working in a small sphere on the dark seafloor. If you watch the videotapes made during dives on the *Alvin*, you begin to get a sense of the inhospitability of the deep ocean as a research site. You can see how strange it is for people to be in a small sphere in a large ocean with little vision through the "fog" of particles floating in the water. The sense of isolation there is palpable. Perhaps when scientists dive on the submarine often enough, it begins to feel as familiar as driving their cars. But for most people most of the time, when they dive to the vent sites, they are surrounded by "nature" and they are a long way from home. Under these circumstances, scientists develop psychological skills, emotion-management techniques, to help make them feel less threatened. Primarily, they bring bits of their surface culture with them to the seafloor to help them sustain a consistent view of themselves and to make the submarine an island of human control in the vast expanse of ocean.[9]

Music is the kind of culture they most commonly use this way. Researchers frequently play tapes to help pass the hours it takes to go to the bottom of the ocean or the boredom that builds up sitting on the seafloor, waiting for experiments to be set up. Like British colonials in Africa having afternoon tea or eating dinner by candlelight in the bush, the scientists listening to and talking about rock or chamber music on the *Alvin* may seem astoundingly oblivious to their surroundings. But they are just hanging on fiercely to their everyday habits.

Jokes are also used by scientists to help normalize the dive situation. One time, when a scientist was trying to grab samples of some worms with a robot arm, he tried to push a crab out of the way. There was a short little "scuffle" as the crab decided to be stubborn about it and try to fight off the submarine. This was an opportunity for the researchers to crack a joke about "crab abuse." The comment was funny in part because the crab's aggressiveness touched on their anxiety about their power *and* powerlessness. The researchers were so much more powerful than the spunky crab that they felt like bullies pushing it around, but at the same time, the scientists could only survive on the seafloor because of the metal shell they had built for themselves. The crab was in its natural habitat, and was fighting to maintain that habitat undisturbed. Joking about crab abuse was also a way to bring current "surface" culture into the deep. There was a lot of media interest at that time in child abuse, so talking about "crab abuse" was a way to absorb the crab's behavior into that cultural frame.

In these and many other ways, scientists bring the culture of their societies to the seafloor to "normalize" the experience of being there. They employ framing devices to mediate between themselves and the deep sea so they can do their work in a routine manner, without being unduly awed or anxious about being at the bottom of the ocean in a tiny sphere.

Researchers who go to sea (whether or not they go down in *Alvin*) also have to develop family arrangements that allow them to leave home for extended periods. They end up organizing lives that resemble those of military families. Some reduce the stresses by marring into the clan (of ocean scientists) and bringing their spouses with them when they go on cruises (or at least having a spouse who will understand when only one can go to sea and the other has to stay at home). Others worry more about their relationships with their kids. They try to convince their children of the importance or romance of the enterprise to reduce resentment about their leaving. These are the only examples of the family arrangements that the scientists I interviewed mentioned to me, but there are probably many more. Even laboratory colleagues may not be aware of all the ways researchers manage the problems of having two lives, one on shore and one at sea. What is important to them (and their funders) is that researchers have techniques for leaving home so they can devote their cruise time to intellectual puzzles. These techniques are part of what make researchers useful members of the scientific labor force.

In spite of their difficulties, these scientists go to sea over and over again. And they get used to the strains on their equipment, their social skills, and their bodies. The end result of this process is that a group of scientists learns to seek (for their own self-esteem or careers) exactly the kind of skills, endurance, and equipment to make deep ocean work productive, and they thereby increase the practical value of the scientific labor pool. Because they cannot (with rare exceptions) go to sea without government funding (and most often they go on government vehicles), they not only develop these skills but they do it in a way that makes their expertise visible to state funders. Some explicitly accept funds on the condition that they make their services available to the Navy when needed. But the rest accept a similar obligation implicitly, and many are quite aware of this fact.

THE CASE OF MILITARY SUBMERSIBLES

The skeptical reader may say that it is all well and good that scientific researchers prepare themselves in so many technical, social, and psychological ways for work at sea, but is there any evidence that it serves the state? The answer is yes. There are some very clear-cut cases. Scientists using vehicles like the *Alvin* (or *Flip* or sleds like Deep Tow or Angus) train themselves in searching for objects on the seafloor, and they have used these skills for the military. The *Alvin*, for example, was designed specifically after the loss of the *Thresher* on April 10, 1963, to facilitate military search (and eventually rescue) missions on the seafloor, and it was used to locate the H-bomb lost off the coast of Spain.

Knowing this history provides some insight into the value to the state of the research done on *Alvin*, something which is difficult to recognize in the content of individual projects. The *Alvin* has been used since its inception for myriad basic research projects. Marine geologists have used it to survey, sample from, and take pictures of spreading centers in the deep ocean in order to further the understanding of plate tectonics. Project FAMOUS (which we will look at more carefully in the next chapter) did a detailed study of an area along the mid-Altantic ridge, using *Alvin* as a major tool. There have also been a flurry of research projects at the deep ocean hydrothermal vents using *Alvin* to study the biology, chemistry, and geology of the vents.[10] Looking at the information gathered during these individual missions on the submarine tells you little about why the government would want to have scientists do their research. But looking at the *skills* scientists have developed using the sub can reveal how, by sponsoring this research, the government has kept a group of scientists skilled in doing deep ocean searches.

While I was doing interviews, the military provided another very clear example of how and why it could become interested in having scientists do work on small submersibles. I heard that the military was planning to make one (or two) of its small, deep ocean submersibles available for scientific research. One version of the story was that the military wanted to put one submersible under the control of one of the major oceanographic institutions; another version had it that the Navy wanted to have two of its vehicles used by scientists in a variety of institutions when they were not needed for military purposes. I asked one of my most reliable informants why the Navy would want to do such a thing. He said that the Navy had a problem with its submersible program. Because it always has new personnel rotating through the program, the Navy ends up training lots of people how to use the submersibles, but none of them uses the vehicles for long. The result is that the Navy simply does not have people in its organization with extensive experience with the equipment. Scientists, on the other hand, want to use submersibles frequently to do their work, and they work hard to develop a high degree of expertise in using them. Since the Navy does not need to use the equipment all the time, it seems sensible enough to let scientists train themselves to be the kind of manpower the Navy cannot easily supply for itself.

The Navy seemed to have strong interest in pursuing this proposal, but it initially received a distinctly lukewarm response from the scientific community. Why? The answer is very revealing of the relationship between the state and science. In spite of the fact that they were desperate for submarine time, researchers knew that as long as the submersibles they were offered as research vehicles were primarily military vehicles, their use could be preempted at any time. In theory, *Alvin* could be preempted, too (and has in fact been preempted on a few occasions), but *Alvin* is only used when a large-scale, strategically sensitive search effort is in progress. The Navy uses these other small, deep submergence vehicles for other searches with less strategic significance, like searches for helicopters lost offshore. So *Alvin* is partially protected from preemption. The subs being offered by the Navy in this case would not be. The scientists also realized that the Navy was not really interested in promoting the progress of oceanography, so they were not about to give the Navy what it wanted without assurance that they were going to get a real opportunity to do research. So the scientists started dickering. They complained that these subs were not as well equipped as *Alvin*, and the Navy proposed to establish an engineering support group for these submarines that would help scientists equip them for cruises.

Why would the Navy go to such lengths to get scientists to use their ships? Why would scientists using government funds be so fussy about the deep submergence vehicles that the government was offering to provide for them? On the one hand, it illustrates the elite status of scientists, and the extent to which scientists insist on being treated as such. It is also testimony to the importance of researchers as a source of skilled manpower for the Navy. Scientists know that they can be fussy, and the Navy knows that its interests are served by trying (at least to some extent) to placate the scientists. In this particular case, another important factor was involved: the call for greater efficiency and fiscal responsibility. The proposed arrangement would promote the military's interests in two important ways. It would get scientists to develop the high-level skills in using the Navy's equipment that the Navy needed, while also making greater use of an expensive piece of machinery. Of course, since the government was going to pay for the use of the submersible in any case, there was no real cost cutting by putting it in the hands of scientists, but at least there could be more extensive use of the ship. It was a deal that the *state* did not want scientists to refuse.

The practical uses of oceanography for the military, fishery service, and other groups inside of the government may make the skills of oceanographers seem too applied to parallel those developed by scientists in other fields. What kinds of skills do people develop when they are doing bioengineering, studying the solar system, or looking at atomic particles, and are they even of remote use to the government? If you focus on the special skills that laboratories develop to advance their work and make it distinguishable from that done at other labs, the answer will seem to be no. The skills they use seem so specialized and remote from practical affairs (at least in most cases) that the relationship between science and politics seems remote at best. But the fundamental skills that specialists in the area share, the ones that are common rather than esoteric, the ones that have many uses rather than one—in fact, exactly the kinds of skills scientists take for granted, just like going to sea—have more potential use (and often many extant uses) for solving practical problems. Knowing how to use telescopes, design payloads with experimental measurement equipment for space, or track objects in the sky are fundamental skills useful to the military. Knowing how to create and modify bacteria may be an esoteric business, but the basic skills have many practical applications with political significance (such as creating bacteria that will eat up oil from oil spills in the ocean). And we have already seen how extant studies in atomic physics changed the Second World War.

So in spite of the unusually direct connections it draws between scientists and strategic government interests, the story about the small submersibles is still particularly useful in isolating the characteristics of scientific research that make it so attractive to the state. While groups within the government may occasionally study the same areas of the natural world as independent scientists or may use some of the same research equipment on an intermittent basis, it is scientists who make these activities the center of their social world. They devote a quite different amount of time and of intellectual energy to these activities. They are (quite patently) specialists with the expertise appropriate for their specialty. They acquire their expertise willingly, and compete with one another to be the best in their fields. The state simply cannot hire and control all the experts they might need, so they contract out to keep at least some of the most vital experts "on call." Scientists, having been targeted as strategically central experts, remain a major reserve labor force. Ironically, of course, most scientists do not notice this. The very skills they carry which may have the clearest political importance are often the most ordinary and routine ones for them, but the potentials of these skills are enormous, particularly when they are so carefully and unceasingly fine tuned.

THE CASE OF OCEAN WASTE DISPOSAL

Some of the skills that the state needs from the labor force of scientists may have only to do with the research *process* (like going to sea or using a sub), but others have to do with the content of a project. The Navy may want advice about what kinds of paint to use on ships, and therefore want people doing research on paints, sea creatures, and seawater. The weather service and fishery people may want experts to tell them more about the El Niño phenomenon (an elevation of ocean temperature that increases the severity of storms, reduces fish populations, at least in usual fishing grounds, and can destroy kelp beds that function as fish nurseries). Many research problems that are funded by government organizations are selected for their salience to clear practical problems.

One example is research related to deep ocean disposal of nuclear wastes sponsored by the Atomic Energy Commission and later the Department of Energy. This type of work started to be considered by the government as early as 1956. We can see this in a letter a private consulting firm sent to the AEC:

> At the present time, the major portion of the radioactive wastes generated within the San Francisco bay area is being transported to sea for disposal by the Navy. The ocean areas that have been used as disposal sites have never been checked for radioactive contamination even though they have been in use for over eight (8) years. No studies have been made on the adequacy of containers and packaging procedures from the viewpoint of the long term contamination control aspects. No criteria has [sic] been established for the selection of an ocean site for waste disposal other than depth (1000 fathoms). It is the opinion of the [Nuclear Research Company] that information on the current structure, marine biological population, and ocean bottom characteristics are factors that should be considered when selecting a site for sea disposal of radioactive wastes.[11]

Apparently the AEC forwarded this letter to a group of scientists, all of whom had been involved in the atomic tests in the Pacific. The AEC presumably wanted some scientists to comment on the *scientific* legitimacy of the proposed work. The scientists, as we can see, seized this opportunity to try to develop their own research program, offering to put their greater or at least more legitimate skills at the disposal of the agency. They tried to show how their more "scientific" research on this subject, using the latest theories and models of the ocean generated by basic research, would be far superior to the program outlined in the original letter. They obviously thought that as long as a problem requiring scientific research was being considered by the AEC, they would mine it for research support. Here is a sample of what they said:

> Two basic questions in the marine sciences are: where, and how fast, and by what mechanism do the ocean waters move? How do the plants and animals of the sea extract from their watery medium the substances needed for their vital processes? Answers to both questions are needed in evaluating the possibilities of disposal of radioactive wastes in the sea.
>
> The first of these questions concerns not only the rapidly moving subsurface waters but also the sluggish waters of the depths. . . . Recent observations suggest that lateral diffusion at all depths is much smaller than hitherto believed, and that vertical diffusion across the thermal strata of the sea is very slow. On the other hand, the water motion at great depths is probably much more rapid than classical theory indicates.
>
> In the biological realm, we may state as a general conclusion that radioactive substances not normally present in appreciable amounts in the sea will be concentrated by marine plants and animals, and that either further concentration or dilution will occur at higher levels in the food chain. Radioactive substances are thus a powerful tool for the study of the way in which marine organisms react with their environment.[12]

The scientists offer their expertise to the AEC, not as seasoned researchers, but rather as masters of the substantive fields related to the problem. They point to the magnitude of the scientific questions raised by dumping policies and to the power of scientific analysis, with its Olympian vision and objectivity, for addressing the problem of nuclear wastes in the ocean.

These authors provide to the agency both a taste of their knowledge about the deep ocean, and a model of their way of approaching the problem of nuclear waste disposal. The latter is particularly interesting. The paragraphs in this proposal are models of scientific dispassion. The fact that recent studies by scientists find that currents at the bottom, where wastes have been or are to be dumped, are much more than anyone expected is not cause for overt alarm. It is enough of a threat to underscore the importance of new research, but it is presented in a nonthreatening manner. Scientific evidence that radioactivity leaked into the ocean might concentrate and move up through the food chain intrigues more than worries these scientists. They are prepared to take this as an opportunity for a kind of natural experiment, providing science with a new means to trace patterns of marine life.[13]

This dispassionate stance not only reflects a dominant norm in the world of science, but also provides a reason why scientists can be trusted as helpmates to the government. Scientific dispassion about findings not only legitimates their results as objective (both in their own social world and in the eyes of the public), but also makes them disavow any explicit policy position of their own. They may (and likely have) positions, but these are given to the government rather than presented in the political arena as the voice of science. Their voices as scientists, to the extent that they enter the political arena, are given to agents of the government to use (to bury or to legitimate and develop policies for the state). So scientists who promise (explicitly or implicitly) to sustain their dispassion in the face of potentially highly politicized questions also show themselves capable of promising skills useful to the state.[14]

The proposal just cited was one designed to establish long-term links between scientists and the AEC, and thus it obscures other ways in which routine research can train scientists to be useful consultants to the government. During the early 1970s, for example, the AEC supported a project to develop a deep sea camera, which could, in theory, provide an inexpensive means for locating and monitoring potential dump sites. The camera was meant to be attached to a weighted "free vehicle" that could go to the seabed, allow the camera to take pictures, dispose of its ballast, and then return to the surface where it could be picked up by a research vessel.[15]

With this project and others in the same program, the AEC was trying to get scientists to solve government problems by improving *scientific* techniques. The scientists involved were not asked to make nuclear-dumping policy their primary concerns, but they were supposed to do work with obvious practical applications.

Even more recently the DOE (heir to the AEC) funded a series of studies on the mobility of organisms from deep water areas (like those considered as dump sites) up through the water column to the near-surface areas important to fisheries. The projects themselves looked just like pieces of basic research, but the scientists used them to advise the program director on the hazards or the lack of hazards in dumping nuclear wastes. By allowing scientists to perform *basic* research to answer applied questions, the DOE allowed the researchers to add to scientific understanding of the deep ocean, making them feel like scientists and not civil servants. But by selecting program-relevant projects to support, the DOE created a group of skilled and informed experts for advising on dumping policy.

In these and similar projects funded throughout the government, scientists are given money to do research that will give them technical and substantive expertise with which to advocate or undercut policy proposals. Through "detached" attention to the natural world, they enter into polemics about policy. The *information* generated by research as well as the skills used in gathering it contribute to the expertise in the scientific labor pool that is managed and monitored by government bureaucrats and used on an intermittent basis for the formation/evaluation of policies or the development of military equipment.

MONITORING THE QUALITY OF WORK DONE BY RESEARCHERS

Monitoring the scientific labor force has two dimensions. It includes the kind of things we have just reviewed: overseeing the development of basic scientific techniques with which scientists can address both their own research and practical problems facing the government. It also includes monitoring and regulating the *quality* of funded projects to insure their scientific legitimacy (and hence to justify agency research budgets). The latter is equally essential to the maintenance of the labor force and is the point at which most soft-money researchers are likely to feel government oversight of their work.

Both basic and applied research projects are monitored by the government for their scientific quality, first, when they are initially up for funding and, later, when they are up for renewal. The government obviously does not want to fund work that is not good. It makes the agency involved look bad to the Congress, and it makes any advice or military system based on that research subject to challenge. In the political arena, the value of science lies in its usefulness in closing political debates by translating them into authoritative technical terms. When the scientific research involved has no scientific status, however, it loses this utility. Particularly where the political issues are tense, the scientific credibility of government proposals needs to be high.

That is why the so-called peer review system in agencies is so valuable. By having peers evaluate the scientific quality of a particular research proposal, or, for that matter, the credibility of an entire direction of research or a particular research technique, it can be given a kind of stamp of approval from the "institution of science," whatever that might be. As we will see in the next chapter, scientists often identify peer review as a measure of scientific authority and autonomy vis-à-vis the state, but peer review is clearly part of the surveillance system by which government assures itself of the quality of its research establishment.

This is not to say that peer review has no role in creating a healthy basic research program in the United States. On the contrary, only when the government can monitor the cutting edge of research and keep a group of scientists working there can it have any hope of assuring that the research it funds will not be outmoded and subsequently useless. It needs a basic research establishment and supports one. Scientists need fundamental skills to be good advisors and problem solvers; they also need to experiment with new techniques so they can generate or at least keep up with the next set of basic techniques that will replace the old ones.

So when NSF funds a project to determine what microbiological processes are involved in supporting the biological communities at the hydrothermal vent sites, they are doing a complicated set of things all at the same time. They are encouraging scientists to work in the submarines and develop new techniques to extract information from the inhospitable environment around these sites. They are also encouraging scientists to promote marine biology by following a promising line of basic research. There are lots of unusual animals at the vent sites, living in unusual conditions for the deep sea, so there may be important biological lessons to be learned there as scientists catalog and try to explain what they can observe and measure there. In promoting this line of basic research, they are also doing more. The government is also betting that the skills extant researchers bring to this site will be adequate to allow them to extract some science from their expeditions. This is a risky business, but agency personnel reduce their risks by reducing agency commitment. The research in the area will be monitored, and subsequent funds will be contingent on successful use of earlier monies. Peer review will be used to control risk.

The surveillance function is just as important for applied as for basic research, even though the system of peer review is usually not enforced or is just mobilized in some truncated fashion by agencies that support applied work. Nonetheless, the advice given by scientists to the state, based on a regimen of applied work, is worthwhile only if it has scientific credibility, so applied projects are also scrutinized for their scientific merits. It turns out that this is a complicated business.[16]

The relationship between the value of scientists as advisors to the government and their skills as researchers in applied work seems simple enough not to deserve much attention. Scientists should be able to do the kinds of work required to solve the problems presented by the government. Period. Of course, there are some wrinkles. The levels of skill required may vary widely. Sometimes the government may simply want researchers of average ability to solve routine problems; other times it may want some of them to develop atomic bombs or work on space weapons systems, which require research at the cutting edge of the sciences. Under some circumstances, it may want scientists simply to define the cutting edge of science so that the quality of work done in the country is generally kept competitive. But whatever they are charged with, scientists need to be appropriately skilled to accomplish their goals.

Unfortunately, it is not that simple. The government more often than not supports research projects that start out being promising enough, but turn out to have fatal flaws. In fact, this is probably the fate of most funded projects. They proceed until their flaws are apparent, at which time they are replaced by new projects (by the same researchers or new ones) that seem to have a better chance of reaching the same goals. The fact that some projects are dropped by researchers before their funders see the flaws only masks the ubiquity of this pattern.

This situation presents problems for government bureaucrats who want to use advice from researchers to decide policy issues. They need research to be authoritative, if they are to argue for policies with it. But they cannot be sure whether the research they commission will do the job. So they end up supporting and discarding research projects to keep their credentials as policy analysts and managers of science as clean as possible.

This pattern can be seen quite clearly in the history of the deep sea camera project mentioned earlier. The AEC supported the project when it seemed promising, and dropped it when its scientific credentials began to be questioned. This research project was well documented in a series of research proposals, reports, and correspondence, which span its entire funding history, from inception to its ultimate rejection by the government. So the story of its rise and fall is easy to read.

At first, the Principal Investigator was able to argue convincingly that this instrument system was so powerful a research tool for studying the deep ocean that researchers could use it to change substantially the scientific understanding of the seafloor. Researchers were finding evidence that contradicted prevailing wisdom about the ocean floor. Fish were found in quantity where there were supposed to be few; a few fish were found in waters thought to be incapable of sustaining life; large sharks were found in waters where they had not been seen before or expected; currents were stronger on the ocean floor than expected; strange, powerful, and transient eddies that no one had anticipated were found to develop in the deep; and the temperature of the water was found suddenly to rise or fall without explanation. In addition, the equipment was being modified to make it a more reliable and useful tool for scanning the seafloor. The research seemed to be a credit to all involved.

But there were problems, too. Few scientific publications had resulted from the project. There were methodological problems in using pictures when their subjects' exact locations could not be determined in advance. The technology was more expensive than hoped. And the data had not been useful in locating safe dumping grounds for wastes.

Unexpected changes in the data made things even worse. Although early pictures documented unexpected levels of marine life, later ones contained nothing at all or just a few crabs or worms. There was a silence from the deep that was just as unpredicted as the waves of information from initial drops. And this too was disarming and provocative.

> In the last year, . . . [b]enthic epifauna photographs between Samoa and Hawaii and San Diego indicate that there are many areas where the fish populations appear to be less than most of our earlier deep photographs. From the [XXX] Expedition it would appear that the macroscopic sea life on the seamounts and guyots in the mid-Pacific mountains and the area between Christmas and Fanning Islands is sparse. [Extract from a proposal for reviewed funding for the free vehicle camera system]

At first these findings were easily rationalized as new surprises that continued to support the need for just the kind of information the cameras could provide. However, they soon began to suggest that the camera system would not reliably bring back useful data. Continued funding, was requested not for using the existing camera system to do more sustained studies, but for refining or expanding the free vehicle system. The claims for science diminished, and the overall character of scientific writing in proposals for this area of research became increasingly simple and mundane.

Abruptly but perhaps not unexpectedly, the happy tone of the correspondence in the file for this project disappeared. Questions began to be raised about the relevance of this work to AEC interests in the deep ocean and about the scientific merit of the research results. The research had been partially funded by NSF in its glory days, but it lost this support, in part because NSF's interest in technological development waned, but also because of the methodological difficulties of using the equipment.

Severe problems with the AEC appeared in 1972, when reviewers of the annual proposal for the Principal Investigator's contract raised questions about the legitimacy of the research. We can surmise the nature of the criticisms by studying a letter written in response to them; it was sent in February 1972 to a marine biologist who was apparently in charge of reviewing this project at the AEC. There seem to have been three major criticisms: one was that few publications had resulted from the project; another was that the camera drops had been too scattered to yield useful data; and the last was that these techniques actually told scientists relatively little about the creatures who arrived in front of the cameras.

The most important problem was that, as the technology was designed, it was impossible to know exactly where the free vehicle with the camera on it would land in the ocean once its was launched; thus it could not be used to study specific and known geological forms on the bottom in a useful way. For photographing fish in the deep the equipment was hard to use for another reason. The pictures could not accurately measure the size of the animals in the pictures, and there was no way to know which local creatures were or were not attracted to the camera, lights, and bait. In the end, it seemed that this system provided lots of data, but no one was quite sure what it was data of. All the careful logs of drops, clock records in the photographs, and measuring sticks placed here or there could not provide adequate context for interpretation. In addition, the "shooting ratio" (to use a film term) was terrible, since the camera could not be selective. And cameras were lost, leaving valuable images and equipment in the ocean.

If careful accounting of the project's scientific merits was necessary, this research was vulnerable. Hence it did not make good bureaucratic sense. The AEC was also under attack at just that time for misrepresenting radiation dangers, so the agency wanted its research program to have more scientific legitimacy. The Principal Investigator understood this and tried to improve the publishing record and present papers about the research at professional meetings, but this was not enough.

It may be that the project was also vulnerable on other counts. It was, for example, producing unwanted results for those hoping to find safe deep-sea dump sites. The abundance of life and larger animals found on the seafloor suggested that radiation leaks could enter the food chain. It was true that some areas had apparently very little life in them, but it was not clear whether some of the animals there had been scared off by the cameras. It was even less clear *why* the patterns of lifelessness appeared as they did in the pictures. These findings may not have bothered high-level officials of the AEC. Some scientists claim that the agency never took the idea of ocean dumping very seriously. Unfortunately, it is hard to evaluate these political factors, in part because this agency was so political in its uses of science that it was sensitive about its activities.[17]

Perhaps more importantly for this particular project, the image of *science* was not well served by the results. New sets of pictures often contained information that contradicted common scientific "knowledge" about the deep ocean but did not provide solid evidence to substantiate any new theory. The camera was a kind of anarchist in the ordered world of science. Science seemed weakened rather than strengthened by the research.[18]

In the face of this small-scale legitimation crisis, a new voice began to speak for the interests of science and the funding agency. This voice was that of the critical reviewer. If the image of science—its mythic ideal and authoritative voice—could not be supported by this research project, then it would be sustained by the critic, and funds would go to projects that could easily pass muster.[19] Using the camera could not gain scientific credibility, so keeping a group of scientists skilled in its use was ludicrous. The AEC could not legitimate any policy decision made on the basis of such questionable data without making itself vulnerable to its many critics. So agency interest in the project died with its scientific legitimacy.[20]

In this example, we can see quite clearly how the funding system linked autonomy, scientific legitimacy, surveillance, and advice. The chief scientist developed the project himself, in part according to his own research agenda. The AEC did not propose this research; he did. At the same time, he was required to write a series of proposals for continued funding of the project that reported to the state about the work. The system of surveillance within the agency kept tabs on the progress of the project, both to assess what kinds of advice the Principal Investigator could provide them about waste disposal and to assess the scientific legitimacy of the work. When the project appeared to have lost scientific respect, the funding was cut. The life history of the project was determined by a combination of scientific autonomy, agency surveillance, and desire for authoritative scientific advice.

Through selective funding like this, the state has maintained its access to the expertise of a pool of scientists able to evaluate policies and lend their voices to the state for conducting political debates. Scientists have been required to attend to their scientific reputations and the scientific evaluation of their projects to keep the authority of science in their voices. When done successfully, this has given them something to trade with the government for funds. It has also made the voice of science available to the state as *its* resource for approaching issues of vital importance to nations and the natural world, from weapons policy to disposal of nuclear waste.

CHAPTER FIVE

Limits on the Autonomy of Soft-Money Scientists

TALK TO MOST successful scientists about their relationship to the government and they will laugh (or scowl) at any suggestion that the government bureaucracy can effectively control anything as large and complex as the scientific research system—much less use it in a systematic fashion to enhance state power. They will agree that government officials, particularly people in the military, make concerted efforts to do so, but they will also remind you that while the Department of Defense is a major funder of research, it is neither the only nor most vital part of the research establishment. They will also probably remind you that the government needs scientific help as much as scientists need government money, and that scientists have been powerful policymakers since World War II.

There is much truth to this point of view, but it also contains a fundamental misunderstanding. Certainly, the scientific research apparatus exists precisely because the state needs scientists to develop and evaluate policies and weapons projects. But it does not follow that scientists are empowered as a result. The successful use of science by the state gives *science* a potential value so great that it cannot be ignored. But *scientists* are rarely given much power. Some enter government service or act as major consultants to policymakers or the military. They feel politically potent, but the most powerful of them usually have to give up research careers to do so. In contrast, active researchers may give advice to government policymakers based on research, but they will not make the political decisions from that advice. Scientific perspectives may be used to develop or legitimate government policies, but scientists will not be direct political initiators or beneficiaries of them. The Research and Development budgets may expand when scientific advice seems strategically vital, but the power of individual scientists will usually not be affected.

Ordinarily, soft-money researchers do not worry much about their political power as long as they do not see themselves as victims of state manipulation (and normally they do not). If they see their science flourishing, that is usually enough for them. They can look around their labs and see no bureaucrats peering over their shoulders, telling them what to do; they see themselves doing research that they have either initiated or volunteered to do; and they understand that in the major science-funding agencies, the National Science Foundation and National Institutes of Health, projects must pass a peer review before being funded. Hence they see little outside direction of their research, and they take this autonomy as a measure of the power of science vis-à-vis the state.

Again, there is much truth to this. The soft-money funding system in the United States was devised in part to keep most research free from direct government oversight. National laboratories exist, but the vast majority of funded research is done elsewhere. Researchers doing basic research on NSF money can easily think of themselves as working wholly within the social world of science. They get money by pleasing scientific reviewers with the scientific merits of their work, and they use their research to serve science, not policy. Interestingly, many scientists doing applied work feel much the same way. They write in their proposals about the applied potentials of their projects, but they think policy issues are really quite marginal to the work they do. They see more discretion than circumscription for scientists in the funding relationship.

But none of this really negates the power of the state to fund research selectively for its own (albeit diverse) purposes. Scientists may initiate many of the research projects they conduct, but most of the projects they propose are rejected. Fred Spiess found the following ratios of proposals to principal investigators for oceanography at the two biggest oceanographic institutions, Scripps (SIO) and Woods Hole (WHOI):

> In 1975 at SIO there were 129 potential principal investigators (P.I.'s) and they submitted 250 proposals. At WHOI that year there were 95 P.I.'s and 260 proposals. Ten years later (1985) there were 152 P.I.'s at SIO submitting 463 proposals while at WHOI there were 110 P.I.'s and 455 proposals. In that ten-year period the SIO ratio of proposals per P.I. rose from 1.3 to 3.0 while for WHOI the change was from 2.7 to 4.1.[1]

Even the most respected scientists have to hustle for grant money, in large part because they have bigger salaries to underwrite and larger research establishments to support and hence need to juggle more projects just to keep their normal activities going. One man put it this way:

> You figure, you know, say [that you are] an established researcher, you know, getting a salary, you know, somewhere around forty or fifty thousand a year or something like that. That means the overhead of the institute will be 65 [percent in addition] on that, depending on the nature of the work. So he's gotta raise twice his salary just to start. And so he's trying to write three proposals a month. There's nobody gonna give you a hundred thousand dollars salary for one proposal, so you've gotta do three proposals at once. You talk with the Navy; you talk with ONR; you talk with NSF; get easy money out of EPA or NOAA or somebody. [Interview with a senior scientist at a major university]

It may make sense at first to think that soft-money researchers initiate their own research projects because they are intellectually autonomous enough to do so, but it is equally true that they must do it in the United States because no one is offering to give them money, and they cannot do research without outside funds. In developing proposals, they do not so much sketch out their research dreams as try to find ways to fit their scientific interests to the priorities of funding agencies, maximizing their chances of success when applying for money. Moreover, they do not make the final decisions about their research futures. Others do that when they decide which proposals (if any) to fund. They cannot even be sure that funding decisions will be based on the relative scientific merit of their proposals. Even at NSF, where the proposals are ranked by peer review, there is still administrative discretion in determining funding priorities. This is even more true at agencies like ONR that cherish their institutional freedom because they have limited forms of peer review. At ONR, for example, administrators try to think about the broader interests of both the Navy and the scientific community, funding projects that are either needed by the government or unlikely to find support in other agencies. Using these criteria, they may put higher on the list of funding priorities projects that peer reviewers did not see as the best science. All this means is that the autonomy of science in its relationship to the federal government is not as great as scientists often think and as U.S. policy analysts have claimed.

Soft-money scientists not only find outsiders shaping their research careers through funding decisions; they also face budgetary constraints on their uses of the research money they are given that limit their discretion in shaping the day-to-day activities of their labs. These are the annoying aspects of doing funded research that soft-money scientists do not like to think about; but by not paying attention to them, scientists ignore governmental constraints on research. All this is not to say that scientists have no autonomy but rather that it is limited and they must struggle to make the most of it. *They* must figure out ways to make their scientific aims realizable within the constraints of the funding system. Getting funds to do scientifically valuable research is not automatic.

The irony is that the very things that scientists take as measures of their autonomy (like the proposals they write) are also indicators of their deep dependence on state money. Even belief in their autonomy itself serves the state as well as scientists. Cold-war ideology that pitted science in an "open society" (the United States) against that in a "closed" one (the Soviet Union) gave scientific autonomy an ideological usefulness in the 1950s that has diminished over time, but not disappeared entirely. The government has not given up the idea that American scientists are and should be intellectually autonomous. There is no reason why they should. Having scientists believe in the ideal of scientific autonomy and try to realize it in their labs has made them better advisors in two ways: more politically disengaged and more creditable as detached advisors. That is why the push by scientists for intellectual control of their labs and long-term control of their research agendas through peer review has not been opposed by the state—just limited by it.

Still, in agencies where peer review is established policy, the actual powers of reviewers are highly circumscribed. First of all, their recommendations are advisory, not determining. Secondly, agency personnel make decisions about who will review proposals, and everyone knows that the selection of scientists to do the reviewing can be as important as the review process itself. There are well-known camps or schools of thought in science that create alliances and enemies. By playing with these social variables, those making choices about who will do a review may not actually determine what the final review will say, but they can dramatically stack the probabilities in one direction or another. Moreover, program personnel at agencies can decide, based on the activities of panels in the past, whom they want to use again on future panels. They can reinvoke critical voices or more supportive ones; they can reinforce or try to ameliorate prejudices. The "rotators" at NSF who work at that agency for only a few years remain so embedded in scientific networks that they can predict with astonishing accuracy how to create their preferred outcomes from review panels. For these reasons as well as others, the administrative power in the system is surprisingly great.[2]

DEGREES OF AUTONOMY IN RESEARCH

How autonomous are American government-supported scientists then? How do they fight for control over their research? What are their problems and what are their strategies for handling them? In what areas do they win or lose in their fight to govern their research?

Less well-established researchers have fewer options for finding research support, and hence are more easily persuaded to trade some autonomy for some regular source of research money. They frequently allow themselves to be recruited for some routine applied research needed by the government (such as coastal surveys or sewage studies). There are two vulnerable groups that fall into this category: veteran researchers who have either failed to gain distinction for their basic research or passed their prime, and new Ph.D.'s without a regular position in a laboratory. The older ones generally turn to applied work when they get tired of scrambling for funds, and want to find a more stable way to make a living. They often have abandoned dreams of further intellectual glory, but they remain committed researchers and want to continue getting

government money. If they are lucky and well connected, they sometimes can find applied programs that actually advance their science while
providing stable resources. Younger ones are usually not so lucky. They
are even more likely to end up doing research with minimal inherent
intellectual relevance for them. They too are committed researchers, and
want to continue laboratory (or shipboard) work. They often turn to applied projects or national laboratores for jobs. The only alternative they
can see is teaching in a nonresearch institution, and many young scientists just do not want to do this. They want careers in research.

> Everyone early to midcareer in my field who has stayed in academia has had a
> very difficult time. . . . Some of them have farmed out into national labs like
> Los Alamos or . . . Oak Ridge. There are two different people that are
> [there]. . . . The national labs support a sector of research and they have to
> work on certain projects that are of specific interest to weapons development
> or energy development or things that, you know, the government wants to
> support like that, or nuclear waste disposal or whatever. But then they have a
> certain amount of leeway to do their own research, too, so that's one thing that
> some people have done. Other people have gone into chemistry . . . although
> right now, with the crisis in the oil companies, that wouldn't really be an op-
> tion, but some people have certainly gone out and tried to get jobs there. But
> I know other people who are just hanging on in academia, in temporary jobs,
> in soft-money jobs, submitting proposals and hoping things will get better. I
> know a lot of people who are quitting and about to quit. There's a lot of attrition
> among this particular group of people, my generation . . . of a period of about
> 5 to 8 years there. [There were lots] of students coming out at the end of the
> 1970s and in the 1980s [who went into ocean research when it was expanding,
> but found it contracting while they were getting degrees]. They've been hang-
> ing on in various ways but a lot of them are bailing out. Some of them are
> finding faculty jobs at institutions that have no name in marine work or that are
> not the most appropriate institutions for them to be at, but they're hard up for
> money. They need a job and so they're going for positions that are, have been,
> more teaching than research, even though they prefer to do research. It's that
> sort of thing. [Interview with a junior researcher at a major university]

For these researchers, the idealization of independent research does not
serve them. They lust for this way of life, but they have to compromise.
They either must forgo some autonomy and take a job at a national lab,
give up research for teaching, or accept research money for a government
program with minimal intellectual promise for them. They often choose
the latter in the hope that they can combine applied funds with money
for basic research, and gradually establish reputations as researchers.
Many are wrong about this; they do not get out on their own. But this

strategy does work for some young researchers, and this encourages others to try it.

Peer review for either tired veterans or new researchers is not a source of power; it is the system by which empowered scientists maintain their control over them. It is a system that often pits politically knowledgeable researchers against young, green ones in the allocation of funds. Hence young researchers often feel particularly victimized by it.

Trying to start a career as an independent research is an act of faith by young scientists. They take the dream of intellectual independence and of research funding based on merit, and they keep fighting to realize it in their lives. Like all acts of faith, it is a risky business. Many simply throw their best ideas out into the public arena, hoping their merit will be noticed, without knowing how to defend them, or for that matter how to maintain control over them.

I think there's enough money available to support interesting research. [Pause] Although good projects are not getting funded these days. . . . [If] you have a good project you simply keep resubmitting it. And finally, one of these times it goes through the hoop and there's enough money to go around to it. And if it's good science, eventually . . . sometimes you submit a proposal . . . like I did a few years ago while I was still a graduate student. I submitted a proposal that was ahead of its time in the fact that the communities just didn't recognize it as the important thing to do, but then shortly after we wrote the proposal, almost everybody started realizing that this was exactly the next thing they ought to be doing, and all these proposals popped up and people started trying to do it right and left. So it got turned down and I was young enough to not realize when you know you've got a good idea you should resubmit it. 'Cause when it didn't get funded I just went, aw, you know. And then let other people run away with it. And . . . that was a big mistake but I learned from that. Now I know. . . . If the reviews are not terrible, then you simply improve it and submit it again. If you're consistently getting terrible reviews, you're still wasting everybody's time and people get hostile toward you, at NSF and out in the scientific community. So that's not to your benefit.

. . . But if it's basically getting pretty good reviews, then there's enough luck and play in the system of reviewing and everything and in how much money there is to go around at a given time and in a given program that . . . you can play the odds a little bit. If you submit it three times, one of those times, if it's basically a good idea, there might be enough money to go around and you'll get funded. . . . Or if in somebody's estimation there's a better idea, then it won't get funded and if you never submit it again . . . somebody who reviewed it might write up a proposal eventually and say, well this never got done. This was a good idea. And they'll get funded. [That's what essentially happened with my proposal.] I mean, it wasn't stolen from me in a straightforward way because I think it was an idea whose time had just about come anyway. But we were just a little ahead of the game and when we submitted it, it wasn't grasped as a good idea to do. Also, the other thing, like I said, I was young then and I didn't realize that I should have taken the same proposal and elected to use another instrument than the one we proposed to use and just resubmitted it with the same scientific rationale but with a different instrument and it probably would have gotten funded then. . . . The instruments have different capabilities and the other instrument probably could have achieved the goals better and taught us more. [Interview with a junior researcher at a major university]

Young scientists can indeed "learn" from these funding problems, if they become good enough at proposal writing and getting proposals through the review procedure to obtain money and protect their research ideas from raids of this sort. If they are tenacious as well as qualified enough, they can end up becoming members of the next generation of successful researchers. But their ability to support their research through the funding system is not exactly something they are *given* by the state (or by senior scientists, for that matter); it is something they fight for every time they present new proposals to the government. And what they get when they achieve this success will be something less than autonomy to run their labs as they see fit.

More seasoned researchers, who know tricks for making projects attractive to an agency, may feel they have some control over their research lives, but they are not autonomous by any stretch of the imagination. They can and do acquire skills in shaping funding decisions. Over the years, some cultivate relationships with program directors, serve on review panels, or even work as rotators at NSF, and they use their connections to try to shape agency funding practices. They know that agency personnel (particularly nonrotators) need to learn about developments in science through scientists themselves, and that program officers trust most the scientists they interact with routinely. Using this leverage, scientists can become intellectual entrepreneurs, pushing for more support of those avenues of research they favor the most. The dependence of long-term funding officials on the success of the scientists they support to justify future budgets, the greater stature of well-known senior scientists vis-à-vis many younger scientists at NSF and ONR, plus the fact that research institutions themselves are more likely to help their senior researchers make deals with Washington than younger researchers all help senior researchers to exercise some power at government agencies, but they do not fundamentally change the fact that scientists must work for their limited autonomy. In fact, they simply delineate kinds of *work* that senior scientists can do to shape government funding decisions.[3]

With ongoing projects, experienced researchers have other advantages in their interactions with Washington. They can use their entrepreneurial skills and ties to other researchers to make their projects appeal to funders. When they ask for renewed funding of an ongoing project, for example, they are asked to list all the publications resulting from the project, the new Ph.D.'s who have derived dissertations from the research, and names of other scientists who have used the work in one fashion or another. Seasoned researchers with large labs are more likely to have the students to get Ph.D.'s, post-docs to influence, collaborators to name, and publication records to mark the success of the research. All these "results of research" are "credited" to the project, but they often reflect the social or managerial advantages of some Principal Investigators.

One can see clearly how managerial skills were used successfully for a while in the development of the deep sea camera described in the last chapter. They helped to keep the project alive, even when its scientific value was being questioned. For one thing, the Principal Investigator used his contacts with other researchers to make his limited funds go a long way. Since he was not given the funds to take a large ship on a major cruise just to test this equipment, he found other scientists who were willing to take the instrument and some of his people with them on their expeditions. At the same time, he farmed out much of the work on the project to people in his stable of graduate students and research assistants, increasing the manpower devoted to the project beyond the group directly funded for it. In this fashion, he increased the quality of the instrument and the amount of sheer intellectual power applied to its early application. His organizational and networking skills were important in making his limited budget and modest project generate results.[4]

An examination of the techniques exercised by well-known senior scientists to attract funders, colleagues, and support personnel to help with their projects reveals exactly what kind of help disadvantaged scientists cannot mobilize for their benefit. They cannot tell program directors what to fund in their fields if they do not know them and have few publications and honors to legitimate their opinions. They do not have extensive networks or large laboratories to stretch their research budgets. And they are less likely to know how to make a sow's ear seem like a silk purse in the prose of the research proposal (or less likely to have reviewers see a silk purse in their proposal) if their name does not catch the attention of reviewers.

Finding money for research, then, is hardly guaranteed in the system, and being able to support oneself as a researcher in the United States is not a simple job. Many scientists are disadvantaged in their search for funds. They have few credentials, little experience with agencies, and little hope of selling their views of science to program officers. Peer review is not a support system for them; it is a hindrance to their ambitions.

And many seek research monies from institutions supporting applied research precisely because they cannot tolerate the strictures of peer review. In this fashion, the government gets the manpower it needs to do less desirable (more tedious and less provocative) research, while reproducing the value hierarchy that elevates autonomous research. Scientists who have less autonomy than they had once hoped for learn to believe that it is not the result of the soft-money funding system but their own fault for not being better at their work.

FUNDING STRATEGIES

Well-established scientists, particularly ones engaged in respected research programs, not only use their persuasion, their proposals, and their entrepreneurial skills to try to get funding agencies to give them the kind of support they want for their research; they also try to limit the power of particular funding sources over their work by the ways they seek and manage funds.

Agency Relations

One obvious way researchers manage their research is by carefully choosing their primary contractor, and getting one that will understand the research agendas of their labs and be committed to supporting them. Some scientists can shop around for support because their work has potential interest to a number of funding sources. The peculiarly decentralized system for allocating research money can be put to their advantage. These scientists can leave a funding source if the price paid for staying with that particular agency is too high, or they can take a bit of money from all of them to support a project. One scientist I interviewed said he was changing away from NSF funding because the insecurity of it was too great. He was tired of continually writing up new proposals that went through elaborate peer review. Often he was funded, but sometimes he was not. It just was not worth it to him to do that much work and yet have to live with that insecurity. Because of his high standing in the field, he had a choice about which sources of funds to pursue, and he found he could get money from a more applied program. The agency that he eventually turned to was delighted to have him part of the research group and made every effort to make him happy with this choice.

Scientists circumventing the power of individual agencies by using multiple funding sources often can gain another level of control over their work. They can piece together projects that no one funder would support. Of course, not all scientists are able to get multiple agencies interested in their work, but those who can have the opportunity to put together research programs that fit their scientific agendas closely. Scientists can also gain greater control of research they support this way. Since no single funder has fiscal control over an entire project, no one can dictate the fate of the research. This gives scientists greater autonomy in their work.[5] Here is an example from a project to develop a new deep ocean research vehicle:

> Well the motivation for this new vehicle actually came out of the Sandia seabed disposal program. And the stimulus came out of X, a biologist here at [Big Lab]. . . . He was very enthusiastic and pushed it, and Sandia actually gave us some money for a study, a design sort of a study, trying to establish what a vehicle should be like and how you build one. And unfortunately they didn't continue it because they couldn't scrape together the equipment money for it. It required a rather major investment in funds, and they weren't able to swing it in their program. . . . So what we did was we came up with sort of composite funding for it. We proposed it as a Sea Grant program to develop the vehicle, and matching funds came out of a foundation at [Big Lab] which is equipment money given for equipment instrumentation dispersal. And then, the accumulator, for example, came out of some ONR money that supported the development of that, and the development of the computer control system came from [inaudible]. So we were able to piece together different parts of the project from different sources. [Interview with senior scientist at a major research institution]

Another strategy scientists use to try to maximize their autonomy in research is participation in large projects. Large projects provide researchers both with opportunities for using expensive technology that the government would not supply for any one laboratory to use, while also allowing them to develop support networks, groups of scientists who put one another on projects and thus help keep a group of labs in business.

> At the same time that [ONR was cut back], NSF went into it. . . . It so far had done most of its oceanography projects on an individual to individual basis. One, two, or three people would write a proposal and get half a million dollars or twenty five thousand dollars. Then you got the IDOE [International Decade of Ocean Exploration], which had seventeen million dollars to give away for big multi-institution projects. The DSDP [Deep Sea Drilling Project] came, which in those days gave away fifteen million dollars. . . . And project FAMOUS [French-American Mid-Ocean Undersea Study] came and so on, and we entered into the era of big projects. And as a result individual grants diminished in importance. There was less money for them and many people got sucked into the big projects who did not go for the small ones. . . . And the institutions were more interested in the big projects because they yield more overflow. You know, you can stick a lot of people into big projects that would never

be able to make it on their own. And so it helps you in protecting the weak brothers that you probably should have [dumped] five years ago but didn't have the guts to do so. That loads big projects down with a lot of dead weight but their return is small. [Interview with a senior scientist at a major university]

The reason that big projects can be used this way currently is that they are required to demonstrate their "community acceptance." Scientists who want to do projects with five or more scientists in them are expected to demonstrate not only the scientific value of their project, but also its community value. To do this, researchers must get testimony from others, making the project public and available for others to join. This does not mean that everyone who wants to participate can join the project, but it does encourage those who cannot get funding on their own to look for large projects to join.[6] The result is probably less dismal than the researcher quoted above suggests. There are a number of researchers who routinely go to sea in large projects organized by others, but they are not dead wood. They help others with their research.

Interestingly, many of the women scientists I interviewed fit into this category. They keep themselves from doing applied work for the government (and able to write some of their own research articles) by doing their own research at sea while they work on others' projects. They function as a support group within the world of science, having their research agendas shaped by the big projects that are assembled by others, but remaining part of the independent community of researchers by accepting this role. There are, of course, many women who have successfully funded their research careers with their own projects, but it is also clear that many women have experienced discrimination when they have tried to go to sea. Male researchers have not, until recently, been used to working with female counterparts, and they have found the idea of women on ships disturbing. It is not surprising, then, to find women who go to sea more likely to be accepted in a service capacity, fitting traditional stereotypes of female behavior. Large projects seem to have facilitated this, helping some women researchers get access to data by accepting this role.

Whatever their strengths and weaknesses, large projects involving many labs, large scientific issues, and big budgets still remain a major means for allocating research funds, particularly in kinds of research that are necessarily expensive no matter how few scientists are involved. Much "blue water" oceanography falls into this category; research far out at sea always requires expensive ship time. Deep ocean work far offshore, since it is potentially the most expensive type of all ocean research, invites cooperative projects of large scale. Not surprisingly, then, a number of ocean researchers have made careers by locating opportunities for large projects. The history of one such piece of research, Project FA-MOUS,[7] shows clearly how the funding of big science shapes the intellectual, technical and social character of scientific research.

> The reason [Project FAMOUS was started] was almost primarily that President Nixon made one of these wonderful deals on a visit to Paris. . . . They couldn't agree on anything else, I think, but they can always agree on scientific cooperation. So they produced the French-American scientific cooperative program. And they were very abashed to find out there were neither any French nor any American scientists interested in collaborating. And as a result there was, for a good class of entrepreneur, obviously some money available, you see, if you would just step in and say I'm willing to work with the French. And some people at Woods Hole got together with some French cronies and out of this was borne a project called FAMOUS, the French-American Mid-Ocean Undersea Study. [Interview with a senior scientist at a major university]

Like many other scientific projects, FAMOUS had its origins in a funding opportunity, but it could not be supported without a serious scientific rationale. Big money had to be connected to big theory, and this was supplied by plate tectonics theory. The researchers proposed to study in geological detail a section of the midocean ridge in the Atlantic, accumulating more concrete evidence about one small part of a spreading center. The new data were meant to shed light on a region where new plate was being formed. In this way, researchers could see more clearly how the earth's crust was being generated and something about how it was moving along the bottom of the ocean.

This was no small intellectual agenda. The development of plate tectonics had been a revolutionary advance in geology and geophysics. For ocean scientists, the work was particularly important because it showed how research on the ocean could contribute to broader scientific theory. Ocean researchers were now the ones who were looking at the portions of the earth's crust where most of the plates were being formed, the mountain ridges running through the world's oceans. The oceans also contained many trenches where older crust was being pushed down into the mantle again.[8]

> The geophysicists had done just about all they could do with an indirect look at [plate tectonics], setting the stage for geologically doing it, documenting it, mapping it [in the ocean]. So there was an emergent need to get closer to plate tectonics than gravity, magnetics, seismics, and the normal geophysical senses. I got lucky, so what began to emerge was the need for what geologists would call a classic field mapping program. The nitty gritty, the nitty gritty. I mean, geophysics is extremely good at looking at the big picture, sensing initial pa-

rameters that constrain the envelope that you're looking at, like $F = MA$. Well, not quite but damn close. Add in friction and a few other things and you know, shhhh; but so geophysicists were sort of getting the first big quantum look. There's these many plates, there's these types of behaviors, you know, there's this type of spreading rate, the poles of rotation, you know, etc., etc., etc. But not quite. You know, there was a lot of well, what about this rock and that rock and that rock, that were all the exceptions to the rules, and it seems like there were more exceptions to the rules than examples of the rule. And that became the slugging period . . . or slogging into plate tectonics and documenting it on a real basis. [Interview with a senior scientist at a major research institution]

To do a careful study of even a small area of the mid-Atlantic ridge was an ambitious enough undertaking that it justified using a number of researchers, a lot of equipment, and a long time at sea. The scope of the project and its novelty made it difficult for funders to say exactly *how* the work should be done, and thus allowed many researchers who wanted to study details of the ridge a great deal of autonomy in setting or following their own research agendas.

For years I worked on the mid-Atlantic ridge. I was one of the very first people who decided that just going back and forth across it was no good. We had to go and stake out a small piece the best we can and do something with some real detail like geologists do. There are two categories of ocean-bottom oceanographers, the ones who were brought up as real geologists and worked that way for a long time—on land, you know. And when you work on geology on land, you are always impressed with how quickly things change from place to place and how just flying across in an airplane and taking a sample at three hundred miles, it's likely to do very little for you. And those of us who got promoted out of that sort of experience into oceanography have tended to feel uncomfortable with the broad brush approach of others who have grown up as oceanographers and never knew any better than to get on a ship from San Diego to Hawaii and stop five times to make an observation. And those of us who came out of more ordinary geology tended to push for these detailed studies, which now have turned out to be key to many ocean processes. We managed to, and got some interesting results that showed that midocean ridges were really different from what we thought they were. [Interview with a senior scientist at a major university]

This man and his counterparts were able to elevate their conventional object of study by having it central to this massive project. They could extend their study on the ridge itself by going to the ridge with many colleagues and an array of new equipment, and at the same time promote the value of what they already knew about the subject.

Other scientists came not to increase their expert knowledge of the ridge but to experiment with expensive, unproven, but potentially valuable equipment for deep sea research. One group in particular had developed techniques for using research submarines that they wanted to test and advance through Project FAMOUS.

> And so people started thinking, what can we actually do with *Alvin* that would be science? Because up to that point it had been used a little bit for biologists but mostly for water tourism. And I think at one point I heard the number. It had been used for something like a hundred dives up to that point. And sixty or seventy out of those had been a first dive of somebody who never dove again, and it seemed a rather useless way of using such spectacular equipment, and difficult too. And so expensive for one thing. Since the *Alvin* had, a couple of weeks before, been lost and retrieved from the bottom of the ocean, it was now being refurbished. There were plans to improve the depth capability and so what do we do to get money to do that? Well, we need a spectacular program.
>
> . . . [I was] an early, very early user of *Alvin* and one who concentrated very early in the game in turning a submersible into a field-mapping device. And that then began after I did my thesis, which ended the day before I sailed for Project FAMOUS. Yeah, just turned in my thesis and sailed . . . on Project FAMOUS. . . . My reason for being involved in Project FAMOUS initially was not that I was a guru—I mean I just got my Ph.D.; I was wet behind the ears. [It] was that I knew how to map. Because I was put into Project FAMOUS almost as a technician. . . . I can remember just being terrified . . . I'm gonna go down and look out of the window, but what does it mean? Now how in the hell am I gonna look out the window and make a statement that's earthshaking to plate tectonicists? The point of using the submarine was to show its value for producing major scientific results. The only way to justify the expense of building and using it was to do so.
>
> . . . That was the very first time that *Alvin* did major science. And of course, as well we know, it was a huge success. Partially because of being along on that, I was offered a post-doctorate, a sabbatical at [the university] here, immediately after this. And I spent a whole year here with all the geological data, everything except the hard rocks, and I had a seminar with geophysics students. And [a colleague] came over for a couple of months. And between those students and the two of us, we published it all in one year. [Interview with a senior scientist at a major research institution]

Only a major project like this could have supported such extensive use of the *Alvin*. And while the research proved that the *Alvin* was an intriguing tool for deep ocean work, it did not immediately establish widespread demand for time on the sub. So the research ended for the scientists experimenting with *Alvin* as it had begun, in speculation about how to secure funds. Once again one group of them thought about how to support the *Alvin* and concluded that more big projects were necessary to support the ship.

> So after all that was over, we were going back to the Azores and we're watching whales going by, saying, thank God this is over. So many months, and we were so dirty. It would have been nice to sleep in a good bed. [A colleague] and I were sitting at a table saying to each other, now what next? And what next was a very big question mark as to how would *Alvin* continue to be funded. It had been funded [only for] this project. We had no further knowledge of support. Seventy people's jobs were at stake. And it was the only deep-diving submarine and the mother ship. I forgot the numbers now, but it was something in the order of two and a half to three million dollars a year minimum they needed to keep this going. They just demonstrated it worked, and there was no money. So we sat around and we said, now what the hell are we gonna do to keep the U.S. in the deep submersible business? Well the only thing that we could do is to come up with another spectacular. Because if we go back to the old mode we won't get any money. [Interview with a senior scientist at a major research institution]

These researchers needed to convince funders that it was worth spending a lot of money to keep U.S. scientists in the deep submarine business, thereby keeping the U.S. capacity for deep ocean searches at a high level. To justify the expense, they had to become major creators of big science projects. They could not afford the luxury of smaller lab studies, so they became prime movers in making big science the major mode for doing deep ocean work. Big projects allowed them an arena for getting their work funded. Other researchers (as the scientist quoted above suggested) may occasionally use large projects to evade the pressures of individualized grant getting, but many others like the *Alvin* users see big projects as the only viable avenues for their research.

Money Manipulation

The ability of scientists to maintain some autonomy in their work depends in large part on above-board practices; they publish, get multiple funding sources, think up exciting large projects, develop a network of friends with whom they routinely cooperate for funds, and use whatever opportunities arise to advise their sponsors about how to spend their budgets. But in addition to this manipulation of their reputations and other social resources as scientists, they can also use money manipulation to try to do the kind of research they want to do. They put together funds from different places earmarked but unneeded for other purposes, and use these monies to pay for things that would be difficult to fund directly, such as new equipment.

The size of this operation, the machine is so [expensive], it's such that I can't launder money. Usually what I do is put together, say, a hundred thousand dollars from various places, just [for] equipment and . . . [I try] various finagling of budgets without it being up front. But something this size I can't. You know, it's almost half my annual budget. I mean, I can cobble together an eighty thousand dollar piece of equipment over a couple of years by having the companies sell it to me over two years and me laundering funds. But two hundred is too much. So I'm gonna have to go in with an equipment proposal. People go in for equipment proposals usually for things I launder because I prefer to do it that way. I prefer just to carry on on my own without making up an elaborate justification for why I want to do something. I write enough proposals. But . . . this is the first equipment proposal I've written. It's a separate category from research proposals with capital. I'll see what happens. [Interview with a senior scientist at a major university]

This kind of money manipulation is one way for researchers to gain some autonomy in the funding system. In doing it, scientists exercise a small power over their research lives to try to realize the idealized role of the scientist: that of an independent thinker and investigator.

Manipulating money to give scientists research advantages is something that institutions (universities or research institutes) engage in as well as individual researchers.

It's not just simply a question of NSF deciding to give people money; it's deals that people do. Some guys are in a better place to do deals than others. The University of Rhode Island down there charges overhead on all its grants, and then the state takes the overhead and gives it back to the registered institutes. But why the hell does NSF have a big overhead when it goes right back in? So if you got a grant, either you or I, you'd get all the money, 'cause the state launders it, you see? So the state negotiates with NSF for its overhead and its senator is there listening, you see? And then they just take the money and give it back to you so somebody, either you or I, will probably get one of these machines. [Interview with a senior scientist at a major university]

Some institutions have even gone directly to Congress for research funds, bypassing the review system altogether.[9] All these efforts at money manipulation have been designed to reduce oversight and control of projects from the government.

As the scientist quoted above indicates, deals are indeed central to the world of funding. Because of this, the former heads of major funding institutions who have dropped research careers to serve in government can still find places for themselves in academia, not as researchers, but as administrators. There they can use their considerable political skills and connections to help make local research agendas appealing to people in the funding agencies. This includes both advising researchers about trends in funding and knowing how monetary deals are struck between universities and the agencies. They can help suggest how the overhead from a project taken by the university can be structured to maximize the fundability of that project.

Given the money manipulation at the institutional level, it should be no surprise to find its equivalent in individual labs, particularly the larger ones that see more of the deals being made. The scientist quoted above runs just such a lab, and feels he does it successfully because he knows what is going on both in science and in the funding agencies. He believes that when he manipulates money to further science, he is doing nothing wrong; he is only using the agencies' monies wisely. He does not take the money for his own personal use; he only exercises a small area of discretion left open to scientists in their relations with funding agencies in order to do exactly what the funding agencies want: create good research projects with important and significant results.

Even though they are dependent on the government for research funds, soft-money scientists do have a range of social, intellectual, and political resources at their disposal to try to do the kind of research they want. They are not really *given* autonomy by the government to follow their research agendas (although they could not exercise any autonomy if the state did not permit it); they must *seek* it, but they are encouraged to seek it. The larger the theories they address with their research, the better their standing as scientists in their fields, and the more relevant their research to the agendas of funding sources, the greater their chance to do what they want. But none of these resources do them any good unless they become "entrepreneurs" and try to extract resources from the government to suit their interests. And most researchers have neither the cultural capital nor the propensity for entrepreneuralism to make the funding system an effective and satisfying source of support for their work.

VULNERABILITY TO STATE CONTROL

In spite of all these efforts to manipulate, move, or evade the funding system, scientists still remain vulnerable. They may begin a project with great excitement (or at least agency commitment of some sort), but as we saw in the case of the camera system, all research projects come to an end, either with a cutoff of funds or abandonment of the project by the research lab itself. As one man I interviewed suggested, he often changed projects because he was bored with them. Then he added he was easily bored—because funding agencies were, too. His statement articulated beautifully the fundamental dependence of scientists on their funding agencies. He maintained that his boredom with his research projects was his own, but acknowledged that his feelings were affected by his social vulnerability.

The same sense of vulnerability was expressed by a researcher looking at seabed disposal of nuclear waste:

That's one thing that's good about Sandia. They don't require that you work for them all the time that they're paying you. I don't know how to put it any other way: *they buy people.* And they would rather pay somebody full time to keep him, even though they don't have full time work, than lose them and run the risk of not being able to get them when they need them. There's a kind of logic to it. I thought it was kind of a waste of money from their point of view but it really isn't because clearly, somebody who's on soft money is going to find other support and then, you know, Sandia might come back in a year or two years and say, look we want you to do this, and sorry, I'm busy. [Interview with a researcher at a minor university]

This quote illustrates nicely the double-edged nature of applied funding. On the one hand, this scientist says that he has been bought by the government; he is their creature and his work is theirs to use as they see fit. At the same time, he sees the program as generous; the researchers working in it are given some time to do their own work. They have some real autonomy built into the system. But this man, at least, is still expected to drop everything to do work that the agency wants done, and this interferes with his "science."

The chief scientist for this cruise [I'm scheduled to go on soon] was here in [New Lab] this fall. There was an international dumping symposium here. And he was here, and while that symposium was going on I got a call from my Technical Program Coordinator at [the agency] who said, okay, I've talked to this guy. I want you to go on this cruise, you know, and help him. They want to do some deep water trapping and so on and so on. The guy's at [New Lab]; go and talk to him. So I did. . . . [There were some problems setting up the cruise, and I did not know for a long time whether I was going or not.] And in the meantime I'm sitting here going, come on, you know, I can't, I mean I can't operate like that. 'Cause I've got a paper submitted for a meeting immediately following the cruise, and you know, either I'm going or I'm not. They finally said go, do it, you know. 'Cause this is where political aspects get involved. It's useful to have me go just because I might be able to arrange for future work. Plus just having an American aboard is of some significance because then there's this international cooperation, plus I can collect—that is, if the [cruise operators] will let me; I haven't heard from them yet—I can collect samples of animals that they catch for our use. See, for radio analysis. [In spite of the problems] I guess I am happy with this program because it's comparatively well funded and I get to work on deep water fish. And as long as they keep paying me to work on deep water fish I'll be happy as a clam. I mean, I get to do this trapping; I get to learn new things. And I mean, they're paying for it. I'd never trapped fish until I went out on [this cruise] and I learned how to do it. And, you know, so they're sort of paying for the development of my skills, which can only be an advantage to me. . . . And it's fun; it is. The disadvantage to it is that I'm not really progressing very far in [my field in science]. [Interview with a researcher at a minor university]

Scientists funded by NSF or NIH do not have so many strings attached to their funds. They are not explicitly "on call" to be mobilized like researchers receiving some military or applied funding of the sort described above. Moreover, NSF and NIH researchers are *meant* to progress in their fields. So they do not share the same sense of vulnerability to the whims of the applied agencies, but they do feel vulnerable to reduced budgets, changes in funding priorities, and concepts of good science embedded in these other agencies. They may not experience a complete loss of control over their careers, but they still experience frustrating constraints.

Actually, one of the curious things that happened when I first came here was I said, well, maybe I don't want to do all this big lab business. Maybe I should just think about doing some modeling, theoretical modeling of tracer distributions in the ocean or something. So I submitted a proposal with another fellow here in this department and a guy over in engineering who is sort of a physical oceanographer type. [We] submitted it to NSF, physical oceanography, to try to get funding to develop some models, some computer models, the goal of which was tell us something about the ocean's circulation based on the distribution of these tracers—measure natural tracers in the ocean. And it was turned down, you know, partly because I'd never gotten any funding, or mostly because I'd never gotten any funding for physical oceanography before, and because I'd never written a paper on modeling, although my Ph.D. thesis had a lot of modeling in it. But that was a long time ago. . . . And the program director there, he kind of said, look, you know, you're good at making these helium isotope measurements in the ocean, and that's been really great and all that stuff, but if you want to do some modeling you really need to do some of it, and write a paper on it before you can even expect to get funded to do more of it. That was sort of the bottom line. Well, if you are a soft-money person, you can't afford to do that. So that's one of the flaws in the system as I see it. It is that there is no breathing room at all, unless you create the padding yourself somehow. . . . So then I submitted another proposal, which was like twice as much money to build this mass spectrometer, and it got funded. So I said, here we go; we're going to build another lab, I guess. So that's what I did. [Interview with a junior scientist at a major university]

Quite clearly in this case, the research agenda of a scientist has been set by NSF. The reason for this is not to save money or force the researcher to do applied work or even to serve science better. It is just an artifact of the system—the way decisions are made. Researchers are expected to show proven capacity for the type of research they want funded so they cannot easily shift the direction of their research. New areas must be opened by new scientists, not ones with extant reputations, and these young scientists are exactly the people with the most difficulty securing funds.

The vulnerability of researchers is both central to the fiscal lives of laboratories and a fact of their lives not likely to change in the future. Government interests in research are better served by making scientists vulnerable to government termination of their work. To the extent that the government makes decisions about whom to fund and not to fund on scientific grounds (through peer review), it is easy to believe that science is being served by the system; but the vulnerability of scientists in this system serves the state more than science. The insecure fiscal life of most labs makes scientists too often grateful for whatever support and autonomy they get. Peer review just makes the voice of science seem more authoritative and less vulnerable to the whims of politicians, while also helping to mollify scientists' bad feelings resulting from their dependence on the state. Scientists can win themselves a bit of discretion in their work by using more than one funding source of manipulating money and thus avoiding undue supervision by the state. But none of these strategies to empower a particular scientist can change the fundamental balance of power in the relationship of scientists to the state. Scientists may be allowed to do some basic research in this system and even use much of their "applied" research to further science, but they are *allowed* to do this because the government policy designed by Vannever Bush said this approach would best serve the state. And it survives because it provides the state with what it needs: a labor force skilled in the measurement and analytic techniques useful (or potentially useful) to the government, and a group of researchers anxious enough (for the most part) to please funders so that they are more than willing to put their analytic skills at the service of the state when called upon to do so.

For many younger and journeyman researchers, the time to advise the state is imminent. They do not feel they can preserve the purity of their research careers by seeking exclusively NSF funds. They look to the Department of Defense and the Department of Energy (among others) to make their research lives either possible or easier. They gain what autonomy they can within applied projects (if they care) and use what techniques they can either to keep publishing (and improving their cultural capital as scientists) or to normalize their advisory work (making themselves contented civil servants with jobs they enjoy doing).

CHAPTER SIX

Technological Dependence of Scientific Researchers

TECHNOLOGY, more than anything else, highlights the fundamental dependence of science on the state. The same instruments that allow scientists to make reputations for themselves, promote their labs, and stimulate scientific debate also tie them to the soft-money funding system. They cannot use advanced technology without large amounts of money, both because labs with advanced instrumentation are too expensive to run without funds, and because advanced instruments are generally too complicated for scientists to make inexpensively for themselves. Thus most independent researchers who want to use them must buy or lease them. The scientists who are capable of building their own instruments for advanced research are really no more free from government control than the others, because they also need money. It turns out that if they want to design new machines, they generally must work on ones with military or economic potential. Prototypes for advanced instruments and vehicles are usually just too expensive to make exclusively for the use of a small group of scientists. In the end, the kinds of intellectual and technical advances scientists can make are shaped by the types of advanced technology that the military and business allow them to make and subsequently make available for general scientific use.

The problem is that innovators in the military and in industry are required to keep much of their work secret. When some admirals decide to make deep ocean search equipment public (and hence make it accessible for nonmilitary scientific research) or some leaders of industry decide there is a market for some measuring device (and hence will make it available for researchers), they give scientists the ability to pursue new avenues of research. But when the military keeps its machinery secret and businessmen decide that a piece of equipment is not worth making or they want to keep a machine to themselves, they are restricting the cognitive capacities of researchers. Thus scientists' work is circumscribed by technological limits imposed by these outsiders—not out of malice or a desire to frustrate scientists, but as a consequence of the structural relationship between scientists and people in the military or industry who develop equipment of potential use to researchers.

The dependence of scientists on military instruments has another consequence. Scientists using military machinery to do their work end up cultivating skills useful to the state. As they excitedly do their science with the latest piece of declassified (and probably downgraded) hardware, scientists train themselves to work with a version of current military equipment. The better they get at using it for research, the more skilled they make themselves as members of the scientific labor force. No one has to tell them to make their research useful to the state. Just learning how to do it well and keeping up their research skills gives them value as part of the reserve labor force.

TECHNOLOGY FOR DEEP OCEAN RESEARCH

Before looking more closely at this pattern, it may be useful to review the list of common equipment used by deep ocean researchers. There are three major types: vehicles for getting personnel and equipment to the deep ocean; data collection equipment for retrieving samples or images of the deep; and laboratory equipment used for data analysis.

The primary vehicle for most deep ocean research is a surface ship of some sort. It may be a chartered fishing boat rather than a vessel that has been assigned to the national research fleet, or it may be a Navy vessel of some sort. This depends a great deal on how it will be used. Fishing vessels are well equipped for doing some kinds of biological sampling using nets. Specialized military equipment or diving teams are well suited for studying difficult-to-reach areas of the ocean. But most deep ocean work is done on ships in a special research fleet that have been filled with scientific instruments that researchers can use.[1]

A primary vehicle is often used in conjunction with at least one other vehicle for transporting instruments and perhaps personnel to the sea-floor or somewhere near it. There are a number of submersibles that will take one or more persons down. Each has its own depth limitations. The most widely used is *Alvin* (see fig. 1), the submersible mentioned in the text a number of times already. This ship is now a popular research vessel because it can go deeper than most other submersibles that scientists can use, and it can carry three people at the same time—two scientists and a pilot. It also has a wide range of attachments that scientists can use for sampling and taking pictures of the features of the deep.[2]

But there are other vehicles that are also extremely popular systems for studying the seafloor: Deep Tow and Angus. These are both towed vehicles (sometimes called sleds) that are designed to be "flown" over the seafloor while making images of its contours (see fig. 2). Both have been used primarily for geological work, although not exclusively. Deep Tow is filled with acoustical equipment, both down-facing and side-scanning sonar, which can give quite detailed images of the deep even when flown relatively high off the seafloor, where it is not likely to hit anything. It is

Figure 1. The submersible, *Alvin*. This is the vehicle used with dramatic success for detailed geological study of the mid-Atlantic ridge during Project FAMOUS. It was later used even more dramatically for the discovery and exploration of hydrothermal vent systems. (Courtesy of Woods Hole Oceanographic Institution)

a delicate instrument package so it must be used this way. In contrast, Angus is a large steel cage that houses primarily still photographic cameras that are set up to take enormous numbers of pictures using strobe lights. This vehicle can and has been towed close to the bottom where it hits rocks routinely without any permanent damage to the equipment. It can provide very detailed pictures of the bottom.[3]

A newer instrument package/vehicle, the Argo-Jason system, is now being designed for Navy search work. An early version of the system was used to find the *Titanic* and study its interior. The Argo is a sled (i.e., like Angus and Deep Tow) that was used to locate the ship, and the Jason (fig. 3) is a small robotlike vehicle designed to be used with the Argo (but in fact used with the *Alvin* for investigating the *Titanic*). Argo is a more delicate and elaborate version of the towed sled, Angus, that has a sophisticated video system that relays "real time" still video images to a surface ship every minute. Jason is meant to be a vehicle that can move down from Argo to study features on the seafloor in more detail, taking pictures and gathering samples of objects that scientists on a surface ship

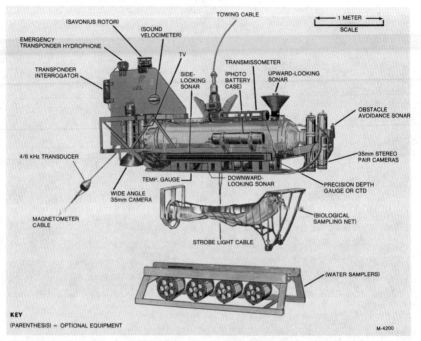

Figure 2. Deep Tow research sled. This vehicle was designed primarily to take advantage of advanced sonar systems for detailed mapping of the seabed before the commercial development of the Sea Beam sonar system. The vehicle was towed over the seafloor to yield composite images of seafloor details. (Courtesy of Marine Physical Laboratory)

can identify with Argo's video cameras. The exact final configuration of Jason has not yet been determined, but whatever it may be, this small vehicle is meant to take advantage of developments in robotics (being a small robot itself and being equipped with an advanced robot arm) to function as a very precise search and sampling system. It will be somewhat like the *Alvin* in its flexibility, but with greater applications because, as an unmanned vehicle, it can be sent into smaller and more dangerous areas of the ocean where a submersible like *Alvin* could not or should not go. The system will normally be used for conducting scientific research, but it can be preempted at any time by the Navy for its own purposes.[4]

Although these are the best known and perhaps most widely used of the research vehicles, there are others in use as well. The RUM series is one. RUMs (Remote Underwater Manipulators) are tanklike vehicles that

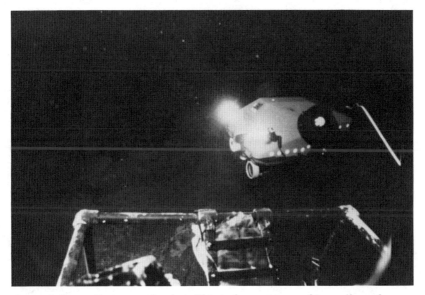

Figure 3. Jason, Jr., research robot. This is the prototype of Jason (from the Argo-Jason system) used in the discovery of the *Titanic*. (Courtesy of Woods Hole Oceanographic Institution)

are tethered to a surface ship and can either be "flown" over the seafloor or put down on the bottom to drive along and do work there. There are, in addition, some sleds designed to be pulled along the bottom, collecting pictures as they go on smooth terrain. Both systems are useful only in smoother areas—the quieter, deeply sedimented areas considered as sites for waste disposal, not the rugged mountains of the spreading centers where much recent basic research in oceanography has been done.[5]

Free vehicles of various sorts have also been devised for deep ocean research (fig. 4). These are relatively small instrument packages that are thrown over the side of a ship. They carry some kind of ballast that sends them down to the bottom, where they take pictures, make current readings, conduct chemical analyses, take sediment cores, trap fish, or whatever. Then the ballast is released, and the instruments begin to rise with the aid of some flotation devices. They are equipped with radio or sonar signals that are triggered when they arrive at the surface so that surface ships can find them, and they can be retrieved. These are generally less expensive systems to use than the larger sleds and RUMs, although they are frequently lost at sea. In addition, they seem to have some practical disadvantages for some kinds of research. They are not very helpful for geologists who want to cover large areas or bring back heavy samples from a particular area they have identified in advance. They are light-weight, good for studying a small area of the seafloor that does not need to be identified precisely beforehand. Thus they are used more by marine biologists searching for samples, marine chemists interested in subtle changes in water chemistry, or physical oceanographers interested in current readings than by geologists, who find the towed vehicles more useful.[6]

In addition to these vehicles, numerous data-collection systems are used by scientists in the deep sea. Some are attached to *Alvin*, free vehicles, or will be attached to Jason. These are water-sampling or testing devices, sediment traps, traps of various sizes for larger animals, temperature probes, pressure cases for storing fish to bring to the surface, thermos bottles for water samples, and endless other devices.

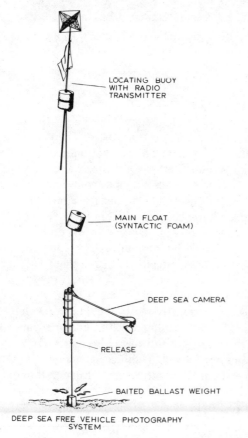

LOCATING BUOY
WITH RADIO
TRANSMITTER

MAIN FLOAT
(SYNTACTIC FOAM)

DEEP SEA CAMERA

RELEASE

BAITED BALLAST WEIGHT

DEEP SEA FREE VEHICLE PHOTOGRAPHY
SYSTEM

Figure 4. Free vehicle system. This version of a free vehicle was used with a variety of different measurement systems to monitor fluctuations in weather patterns in the North Pacific. (Courtesy of the archives of the Scripps Institution of Oceanography Library)

Also, several imaging systems are in use. There are acoustical (sonar) systems, some that scientists use for locating vehicles at sea and making images of the abyssal floor as data; and others, pingers or transponders, which scientists set out on the seafloor to help vehicles locate themselves or some bottom feature precisely by triangulation.[7]

The Sea Beam sonar is one of the acoustic systems frequently mentioned by the researchers I interviewed because it has been so important in recent research on the midocean ridge (fig. 5). It has such a narrow beam that it can take very accurate images of the seafloor from a surface ship. Some advanced acoustical systems are also on Deep Tow, Argo, and other deep ocean vehicles. These include the subsurface systems that can show what is happening under the sediment and the side-scan system that can make images of a broad swath, providing useful survey images for geologists.

Film and video equipment also are used extensively. Two video cameras are available on the *Alvin* for documenting what is seen and how samples are taken. Still and motion picture equipment has been used on towed and free vehicles as well as on *Alvin*, and stereo cameras have been used on both *Alvin* and free vehicles.

A growing area of interest among both marine geologists and biologists is image analysis. These techniques have been developing rapidly in the artificial intelligence community for defense purposes (now that satellites are producing vast quantities of pictures with potential military significance). Ocean researchers have been watching and trying to use techniques as they become available, and the military has been funding them to do this.

Ocean researchers are also learning to use satellites for their studies as this technology is being developed. So far they have used satellites mostly to do worldwide surface surveys of wave and temperature patterns. Scientists have used them to do gravity surveys to produce new images of the seafloor. And there are probably a number of other classified experiments as well, since oceanographers have been involved in the satellite work for the Strategic Defense Initiative.[8]

Beyond these instruments there are, of course, computers: the small computers found in the labs or built into sampling devices; their larger relatives, the mainframes; and even the supercomputers now being made available to researchers. The small computers have revolutionized the design of underwater instruments because they have become small enough to be put in pressure housings and lowered to depth. This means that the instruments used in the deep ocean can now be programmed to do a complex set of activities, providing instantaneous measurements of water chemistry and animal behavior *in situ* of a sort that had not been

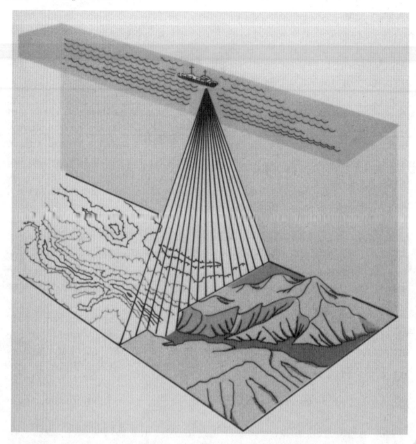

Figure 5. Sea Beam sonar system. This is the type of sonar that was kept secret by the U.S. Navy and then so dramatically unveiled to researchers during Project FAMOUS. (Courtesy of the archives of the Scripps Institution of Oceanography Library)

possible before. The supercomputer also promises to improve the analysis of large-scale survey data such as weather or earthquake data.

These are some of the tools currently used by deep ocean scientists. They are the objects that surround them in their laboratories, that support them or accompany them to sea. They are the machines that facilitate new research and yet also tie scientists to government and business, both because of the expense of using them and because so many of the machines were originally designed to serve the military or other external establishments.

TECHNOLOGICAL DEPENDENCY ON THE MILITARY

The asymmetrical relationship between science and the military is the most dramatic and obvious site and source of technological dependency in scientific research. There we see evidence of the simple inequality in the relationship between science and government: scientists on grants must give their information and equipment to the government when they receive government funding; the government need not reciprocate.

This pattern of technological dependence was immediately visible in the United States at the end of World War II. One of the issues that arose in that period, when science was first being divorced from its wartime submission to the military, was when and how to declassify scientific information and whether to make available for research purposes technologies that scientists had developed or used during the war. Obviously, research would not prove valuable to *science* if it could not be published because it used classified instruments, or, alternatively, if it were conducted with machinery that researchers knew was so downgraded from military models that the results seemed trivial. Still, members of the military were very reluctant to make public the instruments and information they deemed strategically important.

This created a problem in oceanography, for example, when scientists wanted to use sonar to map the ocean floor. Both maps and mapping devices effected the strategic use of submarines, and submarines began to be considered as major links in the defense system. If these ships could hide in seamounts, then should one publicize where seamounts were found? And if the equipment developed to map seamounts could give other countries the ability to make maps similar to ones developed by the U.S. Navy, should it be made generally available for scientific use?

What was much more serious [as a threat to science than scientific involvement in atomic tests in the postwar years] was the Navy's development of ballistic missile submarines, so-called Polaris submarines, because they felt they would have to be concealed. The essence of the Polaris submarines is that you can't find them. That's why they're such a wonderful instrument to maintain the peace. A counterforce strategy doesn't work against Polaris submarines. As long as the submarines cannot be detected or tracked, they're a marvelously stabilizing influence in this terribly dangerous world we live in. But the key word is *not* detected and *not* tracked. The Navy felt that one important aspect of that was to find shallow water spots in the oceans—seamounts, basically— where a submarine could sit on the bottom and not be detected. . . . So the Navy for many years classified bottom soundings. The result of that was that the only people who were taking bottom soundings were the Navy, these secret soundings. The Navy just was incapable of doing anywhere near enough [research]. But the scientists quit doing it, quit surveying the bottom . . . because you couldn't publish the results, couldn't find out what the other guy was doing, couldn't integrate them together. . . . So eventually the Navy decided, after a strong push from me and other people, that it was against their interests, so they unclassified the soundings. . . . I wouldn't argue that we shouldn't classify things that men do, like instruments and devices and techniques. It's undesirable, I think, to do it, but there are obviously some good reasons for doing it. [Interview with a senior scientist at a major university]

This scientist's distinction between classifying scientific information and classifying instruments is instructive, and it helps illuminate better the significance of the reserve labor force of scientists. Controlling the technology, rather than the information produced with machines, means that no one else is going to know more than you are; moreover, no one will know what you are capable of knowing. This implies that research *capacity* more than information is what the government is or should be worried about. The man who made this statement speaks only about technology, not scientific expertise, but a similar logic could apply to scientists. Their research capacities rather than their information are of greatest value to the state.

The view of the sensitivity of technology has persisted and has become an accepted part of the unequal relationship between science and the military. Scientists who work with classified machinery must limit how they let it enter into their science, and those that work with information derived from this machinery must be aware of its meaning to the military. More importantly, what is restricted most are not scientific findings, but the techniques used to generate them. And these techniques are primarily located in the technology and the reserve force of scientists or military people who know how to use them.

The military may be dependent on scientists in many cases to provide the skilled labor force to run their equipment, but they also keep lots of equipment out of the hands of scientists. This fact underscores the subservient position of scientists to the military, the extent to which the intellectual life of scientists is directed by external policies of the state.

Basically what happened was that the U.S. Navy, for totally unconnected reasons, mainly pressures from the suppliers, commercialized many technologies that it had developed for its own purposes like the transponder navigation systems and the big cameras, mainly the navigation system and the Sea Beam, the bathymetric mapping capability. . . . The U.S. Navy has developed acoustic transponders which were basically pingers [for seafloor navigation]. You could send a package down that would emit a ping that would be received by these what we call transponders, which would ping at it again in a different frequency. So if you have accurate clocks, you can clock travel time, use triangulation, and there you are. So they [the Navy] obviously had those for positioning their own instruments to listen for submarines and all that stuff. And the companies who manufacture them, you see, were seeing other applications. This always happens. And so they were demanding that the Navy declassify. You see, what the Navy does then is they truncate; what you get is a degraded version of the technology. You never get state of the art. . . . [A famous inventor and researcher] developed the transponders. But it was a classified thing, well, classified basically, but it was a Deep Tow development. And then the Navy commercialized it, and let the big contract with AMI, American Manufacturing. And then they got greedy and said "Shit, we could see a couple of hundred more." Which they did. And the Sea Beam [mapping system] was the same thing. The Navy for so many reasons decided it wanted extremely accurate . . . it wanted to map the seafloor with the same resolution that we map the land. . . . What happened was then that they started to commercialize a degraded version of this system that wasn't quite so accurate, but accurate enough. . . . We managed to get our hands on a Sea Beam map, first one that was ever released. And then, of course, after the expedition they reclassified them so you couldn't get them. It was unclassified and then some admiral, somewhere, obviously of greater seniority, found out that we had this map and it was at the walls of [Big Lab] and we had to take it down and hand it back in. But we had it for long enough. Then the institutes purchased their own systems through Sperry, so now it is all on the up and up. But oceanography is driven by the interests of the Navy. There are no two ways about it. Satellite navigation systems. These are all degraded versions of Navy operations and everything. [Interview with a senior scientist at a major university]

The essential dependence of science on the military is completely apparent in this situation. What scientists can do is regulated through control of research technology. Moreover, when scientists are given degraded versions of military technology, they work at a technological disadvantage that is structured into the relationship. They learn how to solve problems using this type of machinery, but the military does not give them the best possible problem-solving instruments. No matter how high their social status and sense of autonomy, the power and autonomy of scientists is dramatically undermined by these practices.

Scientists generally contribute to the inequality of their relationship to the military—first, by developing machinery under these conditions, and secondly, by accepting as fundamentally legitimate military interests in secreting technological information (even when it is frustrating).

They can't [give us their maps] because the Navy's reasons are very valid and reasonable, 'cause if they give you a map, it meant they mapped that area, which meant, there's a reason why they mapped the area. So if you go to the Navy and say, we want a map from Fiji, well, they can't even tell you whether they've got one or not because if they say yes or not, then it's yes, we think that's important or absolutely no, we don't have a map, which means we don't have any importance there. So it'd be very easy for someone else to figure out one's strategy by where you map. You could say, oh, so that's the American . . . so they're in a catch twenty-two, they're sort of in a box. But they were nice enough to say, well, if you want to come in, get a security clearance and you can look at the maps. And then we will do selective surveys for you, which they did. In some cases, resurveying and so that, you know [no one would know]. . . . So they were very good and I had no complaints because they provided us with maps when we needed them badly. Before we got the where-withal, which we now have to make our own maps. Now we're out making our own maps. [Interview with a senior scientist at a major research institution]

Scientists with large military contracts have no reason to complain about their treatment by the military, particularly when they are able to get access to restricted information for their own use. But science as an institution has reason to be wary of the asymmetrical relationship between science and the military, since it is structured to keep the knowledge of scientists about certain areas of the world less than that of the military (by limiting its technological capabilities). Since the power of science rests in part of its uses of information technologies, this situation has to be (or should be) disturbing.

The fact that it is not seen by most scientists as disturbing (perhaps reason to be cynical about military support for science but not a basis for protest) is itself instructive. It only makes sense if the scientists involved have accepted their role as an elite reserve labor force. Work with the degraded equipment gives them the capability to work with the more sophisticated machines, if that becomes necessary; helping to design the more complex machines keeps the design capabilities of laboratories up to par; using some of the information from these machines (in a disguised form) can transform some of this work into a resource for science; and using even degraded equipment can give scientists more information than they need for their own purposes.

There can be no quarrel that the research *skills* of scientists in using technology are not degraded in this situation, and cultivation of such skills is what keeps this manpower valuable. There is also no doubt that scientists can use their skills to serve science and enhance their reputations in ways that make the autonomy of laboratories and the scientific enterprise seem uncompromised. But the asymmetry remains and is a structural clue to the social position of soft-money science, and perhaps guarantees that the ocean scientists with the greatest intellectual ambitions will be drawn into working with the military to have access to these stores of equipment and information (or at least drawn to the scientists who already have this kind of access).

The scientific debt to and dependence on the military does not end with the declassification of major data collection technologies; it is also apparent in the use of unclassified military equipment for the design of research equipment, and the adaptation of military models for data analysis to the problems of science. For example, the starlight scope used by the military during the Vietnam war was coupled with video cameras to create cameras that would work at extremely low light levels like those in the deep ocean. One resulting machine has an ASA of 200,000 (the fastest film used by most amateur photographers is only 1,000 ASA).

Adapting this technology for use in the deep ocean has solved, for some situations, one of the more difficult issues in making pictures in the ocean: how to light it. Lighting has been a problem in at least two ways. On the one hand, having adequate amounts of electricity to set off strobe lights has been difficult both in submersibles and unmanned free vehicles with their limited power sources. In addition, setting up the lights has been difficult because of the debris in the ocean that scatters the light. Various ways of positioning cameras relative to lights have been developed to reduce the problem, but using low light is another way to do it. The pictures obtained by these cameras are (as one would expect) extremely grainy and therefore often inadequate for data collection. But they are extremely useful in locating objects for further study (for example the *Titanic* in its recent discovery) and for documenting the use of other equipment at sea.[9]

Another type of technology borrowed from the military is image processing. It was used by NASA and developed for spy satellite pictures. This technology is now being tried by a variety of different researchers in various fashions for improving the usefulness of pictures taken in the deep. Now that it is easy and relatively inexpensive to digitalize photographic images (in fact, one informant said that digitalizing images is the easiest way to store a "frame" of video), image information is in a form that is easily manipulated by computer. Scientists are happy to borrow this kind of innovation from the government, since their use of images is not unlike the use of images for surveillance. The desire among military men for incredible detail in the pictures taken by spy satellites is paralleled by ocean scientists' interest in developing the most detailed pictures of the seafloor. That is why to the extent that the military makes this technology available to scientists (or funds scientists to design it), it finds its way into the laboratories.[10]

Satellite pictures themselves also sometimes suit the needs of both science and military.

The satellite I think has been up there that measures color for about six years [is] just now . . . I think satellites in general are gonna be real useful, and there are satellites that measure elevation of the sea surface very accurately. And I don't know how fine a scale they can get those down to measuring actual waves, but there are satellites that measure wind speed. And so you can get wind data and wind intensity to control whatever happens down there. And then there's these ones that measure temperature and [inaudible] color and there are even satellites now that measure the shape of the seafloor by measuring the gravity so there's images of the whole ocean basin that show features that we didn't even see before. And they come from outer space. It's part of NASA's stuff, I guess. I don't know whose satellite that particular one is, but there was a picture down on the second floor that showed this seafloor topography, global seafloor topography. The advantage of satellites here, for this problem, is that they're the only way you could have instantaneous . . . coverage of the whole ocean, and it allows you to extend these biological measurements in time and space in a way which you can't do. And this also is a remote sensing device. You don't have to be out there. You can measure the carbon flux coming in and it tells you about the biology [inaudible]. I had a mooring out in El Niño, going on in the equatorial Pacific, which is the area that was affected by El Niño most directly. The one that was right on the equator showed pretty much what I had expected in the height of the El Niño period; the carbon flux which reflects the primary productivity was way down. And it was like half of what it should have been, at least. And then I had a mooring that was ten degrees north of the equator and nobody ever seems to go that far away from the equator. . . . And that one had enormous productivity during the height of El Niño. . . . It was to me a total surprise. I always think of El Niño as shutting down the productivity, and all the fish die, and everything about productivity is bad. But here it's marginal to the main area that we normally think of as being affected. There was the huge increase, the highest carbon flux I'd measured in any place in the ocean. And it's all related. I can explain it by . . . essentially doming up of the cold, nutrient rich waters from below just at that one location and in a kind of linear strip across the ocean. And if I had satellite data, it might be possible to confirm that's taking place. [Interview with a senior scientist at a minor university]

This list of debts to the military is incomplete, but helps point to the varied ties between military and science that have affected the development and use of research technologies. The picture that emerges from this accounting is of a science that is dependent in many ways (both conceptually and technically) on the experiments of the military. It also points to the compatibility of interest between the military and scientists in developing the research capacities of scientists. And this, of course, helps make the scientific community an appropriate reserve force for potential military activity.

TECHNOLOGICAL DEPENDENCY ON BUSINESS

Relatively little of the research funds for soft-money oceanography comes from business because businesses (mainly the oil companies) so often hire the researchers they need.[11] Still, there are numerous ties between academic science and business that result from scientific need for research technology. To some extent, scientific innovations are useful to industry and help to define the technological ties between science and business, but much more frequently scientists are dependent on business to provide the kinds of machines that researchers need for their work. There are three ways businesses provide equipment. On the one hand, private companies that have developed equipment under military contract have frequently (as we have already seen) placed pressures on the military to make it available commercially to scientists. In addition, businesses are consumers of research equipment, and their needs often shape major engineering efforts in the country. This too shapes what is available to scientists. And major engineering breakthroughs that have stimulated the growth of new businesses (like the computer business) also affect the work of science.

It is not that industry does not benefit from work of scientists—even ocean scientists. Power companies have frequently used scientists to study water pollution from their nuclear and non-nuclear power plants. Oil companies have used the results of scientific research from both inside their companies and outside to direct offshore drilling operations. And commercial fishing enterprises have used fishery studies, net improvements, and weather information to try to improve their productivity.

The exchange between commercial fishermen and scientists has often been complex. Scientists doing some kinds of dredging and work with nets have used commercial fishing vessels rather than ships from the research fleet for their work. The former are equipped with the most sophisticated hauling machinery available; thus they are good research vessels for these purposes. In return, many fishing vessels have taken advantage of subtle modifications of standard machinery that might increase their competitive advantage. Thus there is a relatively equal relationship between the two groups.

This is not true with the oil companies and other large corporations. These companies have full access to the information generated by scientific research conducted with public funds. Public researchers do not have access to what corporation scientists are learning. Moreover, the technologies developed to make the work of oil companies easier, such as video cameras used for inspection of platforms or pipelines, although made available to scientists, are still designed with the oil companies in mind. They are routinely made for relatively shallow depths and have to be modified for deep ocean work.

The government also sponsors certain kinds of scientific research in part because of its relationship to the oil companies. For example, the Deep Sea Drilling Project,[12] which was so central to the study of spreading centers in the oceans, was also valued highly for its importance to oil companies and others who wanted to know more about the mineral resources on the seafloor. Other research has been stimulated by the interests of oil companies in the seafloor. Some was contracted by the government to try to determine the value of offshore leases. When the Department of the Interior decided to put these plots up to bid, it realized that no one in government knew which were valuable and which were not. The oil companies knew this information and, of course, were more than happy to tell the government what they wanted to bid for the leases. But that seemed unwise, so independent researchers were hired by the government for the job. Thus the oil companies indirectly stimulated additional funding for ocean research.

But most importantly for this chapter, it is the role of businesses as major *producers and consumers of equipment* that is of consequence. Science gains its power from controlling information about the natural world and gains its support from the state for technical as well as intellectual expertise. To the extent that others interfere with its ability to learn about nature (by affecting the design and availability of equipment, for example), science is made vulnerable.

As we saw earlier, businesses with military contracts were the ones who made commercial knockoffs of military equipment available to researchers. The meaning of this to science can be appreciated better when one begins to understand the value of the equipment to science.

We discovered that the Navy had a secret sonar that was to us a James Bond incredible piece of machinery. And [it] was called the SASS system. And the Navy had been mapping the oceans for their own purposes with this sonar. And we didn't . . . know it. And we got the Navy to go into the Project FAMOUS area and map the area with their sonars and the resulting maps were numbing, just numbing to the scientists. . . . I can remember the first time they gathered around a table and looked at them. Almost in just utter disbelief. It was like someone had turned on a light. . . . It was just, you could hear a pin drop; people went, you know [gasp], my God! Look at those maps! The detail was just incredibly high detail. . . . [And] they do it rapidly, twenty knots, wummmp and there was the map. You know, it wasn't like they spent a lot of time. Boom, you know, they have a map to keep you busy for a year. . . . [It was like] the first time anyone had a periscope, I mean a telescope or a microscope. Just my God, look, look, look! You sat there, watching the machine. [Interview with a senior scientist at a major research institution]

Like many other areas of science, ocean research has been affected by the computer revolution. Microcomputers have been made small enough to be built into ocean-going instruments, giving autonomous vehicles new precision and accuracy, making possible new kinds of remotely operated vehicles, and otherwise making some of the new directions in robotics applicable to ocean instrumentation.

Older instruments were updated by the introduction of the computer. From the following examples we can see how computers can make instruments more precise.

I think a lot of that probably turned around when [an engineer] came who was more into microprocessor-based systems. We realized we would have a lot more control of our instrumentation while it was away from the ship as a free vehicle. And this guy came from General Dynamics almost on a sabbatical. He was interested in pushing out [his horizons]. For him [it entailed] a dramatic pay cut for one thing, but it was just something he wanted to do. He came here and worked for a couple of years developing some remote sensing systems. . . . And I think [the Principal Investigator] started realizing the possibility of more sophisticated systems, as did we all. That was probably the major turning. . . . But even at that time standard, we were pretty primitive in terms of [our instruments'] electronic components. . . . (In fact, we're still stuck with some of those systems just because they work, and, you know, they're reliable but they're not necessarily the best. We don't always have money to change toward the best. . . .)

[The engineer] was hired to build a fish trap respirometer that would catch a fish and then close the door so the fish is trapped, and it measures oxygen absorption rates and takes some water samples. Right, it had a series of sensors along the side of this trap so that you could watch movement in the trap, and see the fish swimming around. . . . It had infrared-emitting light sources and photodetectors, and they're all in parallel planes. . . . [When a fish] interrupted the beam, the electronics picked that up as being an active plane, and it's recorded that way. And so when you get the data back you see which planes were active and which weren't. . . . And, anyway, that required some intelligence on the part of the electronics to go into that system. And that was the first system that really had the microprocessor as the control. It was a little more sophisticated than, well, it was a lot more sophisticated than prior to that. Basically, what we had prior to that just turned things on and off at preset times. It couldn't make any decisions. . . .

[Before the first trap] they'd have the submarine actually pick up an animal and put it into a chamber and actually close the chamber. But you had to have some manipulators to get the animal in there. This was the first trap that didn't need somebody down there to put an animal in the chamber. It had bait. And once a fish was lured into the trap, the door was closed so the fish couldn't swim out. . . .

[The computer-based trap] was an idea we could base future systems on. We have another trap now, another respirometer that uses infrared detectors to catch these anthropods, these crustaceans, designed in the same way with infrared detectors. But the traps are only this big. [Interview with two researchers at a major research institution]

Improvement in the accuracy of these traps, using a sensing system to signal when a fish was there to study, made this equipment much more useful for science. It is not that earlier equipment could no longer be used (as the quote makes clear); it is just that this new system was more efficient, since it did not collect data until there was some reason to do so. The equipment was more closely modeled on the behavior of the scientist in the lab, without requiring that the scientist be present at the bottom of the ocean.

Other kinds of more radically novel equipment have also been developed for use in the deep ocean. Until recently certain kinds of small and very mobile ROVs, or remotely operated vehicles (translation: underwater robots), were developed for the oil companies to use in shallow water for inspection of oil pipelines and drilling equipment. These have been modified for work in deeper ocean, and elements of their design have been copied and modified to fit the purposes of scientists. These "robots" work on small computers and programs designed to use them in the ocean. Thus their adaptation to scientific research in the ocean comes from a dual legacy: from the computer industry and from the industries that supply underwater equipment to the oil companies.[13]

The instruments described so far are data-collection systems used to collect samples of some sort and/or to make some basic measurements. There is, in addition to these, an array of laboratory instruments these scientists use for data analysis in their labs that also have been produced in whole or part to serve other markets.

Three of the engineers I interviewed in labs described the design of research instruments for scientists as an assembly process. They spend hours going through catalogues looking for the parts they need, or trying to find an inexpensive version of some machinery they know exists. One described himself as a shopper during much of his day at work. He can see himself this way because what he "makes" in the lab are really modified versions of existing machinery made with prefabricated parts. In this small way, scientific equipment owes its existence to bits and pieces made for other (and primarily industrial) purposes.

Science also owes much of its equipment to commercial laboratories, which constitute a major market for laboratory equipment. The machines they use tend to be commercially available. This includes some of the larger and more expensive pieces that ocean scientists have (or wish to have) in their labs.

> Research is not a driving force behind technology in analytical chemistry. The driving force in analytical chemistry are the boys that make millions of measurements: in clinics, the economic geology companies, pollution boys, the solid state, electronics industry, the really serious followers of contamination and purity of materials. So these guys are the guys, that's the market. I'm not a market. I'll buy one. These guys'll buy five, twenty, a hundred depending on their scale. And so we are basically sitting on the fringes, watching developments in analytical chemistry which are driven by the requirements of others. . . . It used to be clinical medicine but what's happening now is [it's] becoming solid state. The electronics industry is the big driver. They have a lot more money. . . . So we're sitting watching this and saying [to our funders] if you take a plunge here—even though the instrument is not tailored to us, it's got enough intrinsic flexibility, or you can deal with the manufacturer or whatever, because they are always looking for applications—it's gonna be an advance. So that's how we are. . . . We're not in a position to develop our own analytical instrumentation, you know, on the scale I'm talking about. [Interview with a senior scientist at a major university]

There is a variety of ways in which the equipment used by ocean scientists arrives at their laboratories predesigned to fit the needs of business. They cannot afford to make their own equipment from scratch. Many do not have access to engineers or know enough engineering to really do this, and they are not enough of a market themselves to encourage industry to produce new machines just for their benefit. As we have seen, some small companies design research instruments primarily for scientists, but they are always looking for outside buyers and ways to use prefabricated pieces on their own machines. Because of their small numbers, then, scientists do not have the wherewithal (the economic or political clout) to direct technological innovation in ways to best serve their laboratories. The very sophistication in research instrumentation puts scientists into a dependent position.

Once again there is an asymmetrical relationship between scientists and outsiders based on technology. Businesses get the machines they need, but they do not necessarily produce the ones needed by scientists. Moreover, major corporations (such as the oil companies) are much like the military in relation to publicly funded researchers. They are able to learn what soft-money scientists are doing, but they do not reciprocate. Businesses and the state together set the technological agenda and keep that agenda weighted to serve their interests above that of scientists. Once again the autonomy of science, its power to define its own goals, is limited.

Even when scientist *do* design their own equipment, as in the case of the laboratories that specialize in technological innovation, there is still a high degree of control by the state. As we saw in the last chapter, funders have particular agendas for technological innovation that limit the directions in which scientists can be funded for innovation. And the expenses of developing prototypes for new instruments are often even higher than the expenses of buying ready-made machines, so the fiscal dependence of innovators is also marked. Thus, when scientists put together the machines and techniques that they will use for their research, they are working with a repertoire clearly circumscribed by outsiders to the social world of science. And when they use these techniques to do their work, they often are gaining expertise with machinery that resembles what people in the military and business are using. So in the course of pursuing their own research goals, scientists are training themselves as experts in techniques designed originally to be useful to business and the military, not to science.

CHAPTER SEVEN

Techniques and Status in Scientific Laboratories

STUDYING the fate of technology in scientific labs provides another interesting way of assessing state-science relations because it reveals how daily life within labs is shaped by these relations. Much of what people do when they are engaged in research is to use technologies in order to experiment with natural objects. The techniques used in laboratories and the scientific problems addressed with them are the central identifying characteristics of individual laboratories. They determine how laboratories are placed (hierarchically as well as substantively) in the social worlds of science. Yet, for all its importance, the use of techniques in laboratories is a degraded activity, hidden within the laboratory. The very skills that help distinguish one lab's approach from another are developed and sustained by support personnel and machines that are culturally masked from public view, and are publicly attributed to the mind and will of the chief scientist.[1]

This invisible (I exaggerate slightly) world of activity is hidden in part because competing scientists are sometimes concerned about revealing details of their techniques to others before publishing results. Even in publications scientists leave out a surprising amount of detail about their techniques. This is in part because of journal limits on space, but it is more because of the desire of scientists to control techniques to gain priority in discoveries.

Still, the desire to keep new techniques somewhat veiled does not adequately account for the relative invisibility of technicians and techniques in labs. This has a long history and stems in part from the seventeenth- and eighteenth-century tradition of gentlemen scientists hiring technicians as a kind of servant.[2] The nineteenth-century dissociation of scientists from engineers reinforced the downgrading of those who work with machines relative to those who "think" about nature.[3] And in his formulation of U.S. science policy after World War II, Vannevar Bush reiterated the greater importance of scientists to technological innovators in keeping America strong. In all these manifestations, technicians have been defined as subservient to scientists, and this has had practical consequences.[4]

Most chief scientists are not quite so sure that technicians are merely followers and not creators of science. They are often the first to say that instruments are essential to their research and that their technicians are really the ones who know how to make the machines work well enough to do some science. Most firmly believe the old notion that research instruments extend human capacities (like X-rays or microscopes), allowing researchers to see or measure things not accessible to them before. They have no illusion that thought entirely dominates the process of research because they know they cannot think about what they cannot study. They know that having the right instruments and people to manipulate them can make all the difference. They also recognize that one major way chief scientists define their perspectives on common research problems is through the selection of instruments and the techniques for using them. In all these ways, they identify instrumentation and technicians as central to their research.[5]

Many oceanographers also contend that recent history in their field (as much as in physics and biology) has been radically altered by the development of new techniques.

> As of today there are not really very good maps of most of the oceans, which you are probably aware of. What we had at the time [that we began Project FAMOUS] was this relief map, which you may have seen which was made by Bruce Heezen, that has the seafloor in relief patterns, in contours. And people at [Bluc Lab]—I wasn't part of the original team so I didn't have a say in this—they chose the one [spot to conduct the research] that looked the most like a midocean ridge should look. They found it just west of the Azores. It turned out it looked the most like a midocean ridge because Bruce Heezen, having had no data whatsoever for that area at all, had just filled it in with the way he thought a midocean ridge would look. When we went there [in the *Alvin*] it didn't look that way at all. But by that time we were committed. And they were very lucky because at that moment the Navy released a new survey system, which you no doubt know, which has now become the Sea Beam. They had a much fancier version and they declared that they would be willing to do an oceanographic survey for us. And that was the first time that any one of us had seen a map of the ocean floor that was accurate horizontally to about ten or twenty meters. [Interview with a senior researcher at a major research institution]

In this case and many others, marine scientists have been able to make new and more detailed models of nature by trying out new instruments for their research. This has allowed them to define new intellectual agendas and devise innovative research strategies.

One of the most interesting things about this scientist's particular way of describing the advance for oceanography is that he leaves out entirely any discussion of the technicians who made the new equipment work. When he "writes" the history of the discipline, technicians are invisible.

What makes them so unnecessary for telling this story? Why are they taken for granted? There are many elements to it, but one is important to keep in mind. Technicians are "strangers" to science, precisely in Simmel's sense of that term.[6] They are in the world of science, but they are marginal to it. To the extent that they cannot enter that world as independent researchers, they cannot be full members of that world. Because of their dependence on their Principal Investigators, their participation in the world of science is highly contingent; they must please their bosses and help them find funds. As carriers of the day-to-day technical skills of laboratories, they are also tied in quite a different way to the world of government and the military. Often it is *their* skills and training (in the guise of the mind of Principal Investigators) that researchers make available within the scientific labor force. Thus, because of both their dependence and skills, technicians are loci for that difficult and often masked relationship between science and the state. In many ways, then, they work at the point where the "other" (government as well as engineering) enters into science.

The fact that machines and their builder-tenders are central to the relationship of science to the state is made clearer by one of the ironies of the situation. Although most technicians (even technological innovators) in labs have little chance of getting grant money except through their bosses, the Principal Investigators who are themselves technological innovators have access to more grant money than their counterparts doing basic or applied research. The reason is that the government (mainly the military) needs technological innovators more than basic researchers, and it expects those with scientific credentials to be superior to those without them (because of the Bush report and the world of cultural assumptions it embodied). Hence scientist innovators are often very "wealthy" members of the scientific community. But their association with machines still reduces their stature in the scientific community. They have lower prestige than colleagues doing successful basic research, so the hierarchical relation of technicians to "real" researchers (and of those who serve the government to those who primarily serve science) is reproduced.

TECHNOLOGICAL SKILLS AND LABORATORY SIGNATURES

What is the relationship between scientists and machinery in the laboratory? What is going on when scientists channel their work through machines, and learn to think about results by taking into account the limits and capacities of their technology? How do the identities of laboratories get tied to their research techniques? And what are the results of this for the competition among scientists and the relationship between science and the state?

The centrality of technicians and their machines to the life of labs is clear enough. As Bruno Latour has pointed out, when you enter scientific labs, you are not confronted by people thinking as much as you are faced with people working at machines—some of them typewriters and micro-computers for word processing, but most of them machines for data ma-nipulation.[7] These people and machines accomplish the routine work of the lab, and they are the repositories of the technological skills that make research projects work. Their contribution to the research process is par-tially masked by prejudice against technology and those who use it, but what they are able to do together constitutes the capacity of the labora-tory to conduct its experiments and report them to the world.

It is easy to think of science as a purely intellectual activity, and then conclude that scientific skills are mainly conceptual/intellectual ones. But that is a mistake. A large portion of the skills researchers exercise in lab-oratories involve some form of instrumentation. They employ machinery to move, transform, measure, or otherwise manipulate some kind of sam-ple taken from the natural world. It is the combination of people and machines, organized to accomplish certain analytic processes, that con-stitute the techniques of the laboratory. And it is the development and use of these techniques that give scientists their sets of useful skills. Many of these skills are useful for solving both intellectual and practical problems. They may be designed to favor one over the other, but to the extent that they define the capacities of researchers within the laboratory to solve problems (defining both the limits and abilities of this group), they delineate the value of the research to both the state and the insti-tution of science itself.

Different laboratories each have their own sets of techniques that they apply to a limited range of scientific problems. The range of problems and the techniques used to address them change (usually gradually) over time, but at any given moment they determine the analytic capacities of the laboratory. Each laboratory is known within the world of science for its own particular set of techniques. That is why I speak of them as "sig-natures." Signatures are the identifying work techniques of a lab, consti-tuted from a complicated system of relations between laboratory person-nel and machinery, and designed to embody problem-solving strategies for research.

Jacques Ellul argues that a kind of obsession with technique has de-veloped in modern societies (and I would add, their science), which re-sults from work with machines. He argues that a fundamental alienation of the human spirit is involved. As people begin to try to make their activities more machinelike so they can work most effectively with ma-chines, technique becomes fetishized and human qualities are sacrificed. Ellul exaggerates these effects (for effect), but he does point to a melding of the human and inanimate that exists where people routinely use machines, including scientific laboratories. Thus he provides a starting point for considering exactly what scientists are doing when they use machines in research—how they cognitively include their equipment into their thinking.[8]

But David Sudnow, in his book, *Ways of the Hand*, provides an even better model than Ellul's for understsanding the social psychology of technique.[9] The usefulness of the model may not be immediately apparent to students of science because this book is substantively about playing jazz piano, but in it, Sudnow describes how people learn to improvise on an instrument. One can use his analysis to consider how people in other areas learn to think and act in concert (so to speak) with other kinds of instruments, including ones for making scientific analyses. Sudnow himself invites this kind of extrapolation when he compares the process of jazz improvisation to writing on a typewriter or driving a car. In all these cases, he argues, manipulation of equipment is most successful and fluid when it becomes semiconscious, almost automatic. Trying to think about how you shift gears can almost destroy your ability to do it. There is a kind of kinesthetic logic to the touch of hands on machines that "takes over" when machine use is going well. This is when musicians begin to improvise well or drivers lose consciousness of their actions at the wheel and simply feel themselves move as vehicles on the road.

There is a kind of alienation involved in this. But it is only superficially caught by Ellul's concept of technique.[10] There is a submission to a limited range of human senses and capacities that is addressed by the machinery. There is a concentration of energies in a certain direction achieved through the sacrifice of other faculties. Researchers at a microscope concentrate on the sense impressions gathered from the microscope rather than on what they could smell or hear. At the same time, the microscope narrows the field of vision to a small area in order to make it larger. Both machine and its user contribute to a kind of "tunnel vision." Something similar happens in a research submarine as well, although it is less obvious at first. In the dark and murky waters near the seafloor in most areas of the deep ocean, scientists can see very little. There are small pools of light created by the submarine's exterior floodlights, but even they do not pierce the murk very effectively. Scientists in such a submarine concentrate attentively on what they can see, what their equipment is doing, what kinds of readings they are getting from it, and where the submarine is or is not going. The result is that the researchers get a relatively intimate look at very small portions of the seafloor, blind to what goes on beyond their field of vision, the capacities of their instruments, and the input from the senses that they are not treating as significant.

In these situations, machines and their tenders become (at least cognitively) a corporate entity. The end of one and the beginning of the other may be physically obvious; one can see where the hands hit the typewriter. But the activity of making music, conducting an experiment, or writing a line of prose is the act of the corporate entity, not its constituent parts. This entity is a social one in that it is directed to function in the social world; it is also a technological one in that it cannot function without the machinery being in good shape. It is also less than wholly mechanical and less than wholly self-conscious. The Cartesian self and Ellul's embodied technique are extreme cases on a continuum; scientific researchers at work with their instruments lie somewhere in between.[11]

Research is done on a day-to-day basis through thousands of routine measurements, displays of technical dexterity, and unself-conscious attempts to see, feel, or move nature through technology. Laboratory workers begin to think like machines in the sense that they easily submit their activities and understanding of these activities to the restricted properties of the equipment. They do not think about what a microscope is doing to their perception as much as they "see" through their understanding of the microscope and what it does. Much of this understanding is never verbalized; it comes from experience with the machines themselves. Once they can think "through" their instruments, scientists are able to design more successful research projects because they can better anticipate the results. They can then improvise on standard research practices to find their own individual "signatures."[12]

> What I've been doing to count the colonies [of bacteria] and the distributions of them in tubes is to count them with an image analyzer. Part of that was just to develop the technology to see what the image analyzer was good for, and it's a really sophisticated piece of equipment. It's fantastic. . . . [The chief scientist in the lab] has this device which is a temperature gradient block; it sets up a temperature gradient from zero to fifty degrees in pressure chambers. So you put a bunch of bacterial cells in a long glass tube in agar, which is a solidifying agent, and so they're distributed along the whole length of the tube. You shove them into a temperature gradient from zero to fifty degrees and pressurize them. And I had eight of these chambers and you put them at all different pressures at the same gradients; then you leave them for a period of time, take them out, and then you look to see in what temperature range the colonies formed. Well, that's fine and [the chief scientist] had used this device for a long time. You'd mark the extremes and then you'd find that at higher and higher pressures the extremes get narrower and narrower. That's fine, but there were some subtleties going on that were really hard to resolve. Like there were areas where the numbers of colonies were different in the tubes, and so I wanted to be able to quantify the numbers of colonies and it was really hard. I

mean [if] you sit at a microscope and try to count these things you go nuts. So I sat down with [the lab technician] who is our programmer, and I said, you know, we have this image analyzer and the two of us worked out a program. Basically, I was dictating the criteria for the program and he was writing the code, and it works great. It's really a nice program. In fact I just finished the paper on it. . . . The image analyzer looks through a microscope and it photographs a small area of the tube; then it transforms it to high contrast, and then it circles and counts each of the colonies. . . . You have to have something that's fairly uniform and the world isn't uniform. . . . The only reason I had any success, I think, is because the system I was dealing with was fairly uniform, and the processing that I was doing were relative measurements so that because it was uniform I could say, "Well, even though I'm not counting a hundred percent of what's in this tube, I'm doing the same thing all along the tube so that the relative numbers are going to be valid even if the absolute values aren't." And the image analyzer was great for doing that. It generated some data that it would have taken me ages to be able to get by eye. [Interview with a post-doc at a major research institution]

The importance of being able to "see through" machinery is also central to their design. One image-processing expert I interviewed said that one function of image processing is to correct for distortions in the images made by machines. When I asked him what this meant, he said that images produced by an electron microscope, for example, would not "look like" the objects under study if corrections were not put into their programs. One could argue about whether the resulting images were more or less "real" than the ones produced more directly from the microscope. But what is clear here is that the "corrections" made through image processing made it easier for researchers to "see through" the equipment, that is, to act as though they were seeing these objects without aid. The machine is made less visible to the eyes of the researcher. The resulting instrument is more of a "black box" in Latour's sense. How it works is made invisible. What is important then is how to use it. The machine is defined by how it is incorporated into the culture of the laboratory; its technical history is erased, and its internal functioning is made irrelevant to its daily use.[13]

A combination of this tendency to design or customize machines to facilitate seeing through them plus the development of laboratory skills at working through instruments helps to promote a kind of symbiotic relationship between laboratory workers and their machines that gives laboratories their peculiar characteristics. Each laboratory has its own corporate character built from these relationships; this is its "signature." In the social world of scientists, the research characteristics of individual scientists, are, of course, equated with the signatures of their laboratories. But while chief scientists may be given the nominal authority for the lab's activities, it is the *laboratory*, not the isolated scientist, that creates a distinctive pattern of scientific research. The broad characteristics of the signature may be designed by the Principal Investigator, but many of its details are not, and often they are ones that make the laboratory function. Thus machines and personnel interact to produce a coordinated and specific array of research activities, constituting a signature.[14]

The social world of scientists revolves around the development of and changes in signatures, contests between "schools" associated with signatures, and cooperative uses of the skills (or signatures) of different labs for large projects. Signatures are strategically important for labs both in establishing the identity of researchers to funders and in making claims about scientific territory. How this works we will see in more detail in the next two chapters. What is more salient to this chapter is how technique, specifically the signatory techniques of labs, enters into patterns of cooperation and competition in science.

This issue was immediately important at the end of the war. One scientist put it this way:

> [Scientists] were quite willing to help the war effort any way they could, but it turns out that the only thing that scientists can really do is the thing that they know how to do, not the things that you think of for them to do but things that they think of for themselves to do because they know how to do it. . . . It's basically because he [the scientist] has certain tools at his command, certain technologies or techniques, which work in certain ways but not in others. Unless you're very familiar with his technology and his methods, it's awfully hard to design a project for him to do which he can really do.[15]

Signatures give laboratories peculiar sets of capacities and incapacities for research. They make laboratories oddly inflexible creatures, particularly in relationship to funding agencies. On the one hand, if the scientist quoted above is correct, signatures contribute to an autonomy of scientists that prevents funding agencies from effectively dictating their work. (Obviously there is some ideology involved here, but the point is still worth making.) On the other hand, the existence of signatures leads to a conservatism among funders. Funding agencies will support laboratories only when they propose to do the type of work for which they are known. This hampers researchers and is a constraint on their freedom.

Laboratory signatures are also at the heart of competition among scientists. Laboratories develop distinctive approaches to the problems in their field. These approaches are compared, and the status of laboratories in the field depends to a large extent on how they make out in these comparisons. For this reason, laboratories headed by a particular scientist will often take a "posture" of advocating some set of techniques or an approach to a problem that the lab has been able to use successfully. A great deal of time, money, and creativity go into shaping these positions, so scientists are often reluctant to give them up. There are always stories in science of researchers who refused to accept findings by others that undermined their own research. This obstinacy makes more sense given the social and technical as well as intellectual aspects of these stances. They bind people, machines, and theories into elaborate constellations. The price of changing them is often very high monetarily, socially, and cognitively.[16]

Of course, this social rigidity in some labs and the conservatism of funding agencies do not mean that laboratories cannot or do not change their techniques. They frequently do as a strategy for improving their positions in their fields or vis-à-vis their funders. Changes by any one lab may lead to changes in others as they strategically look to gain a competitive advantage. But all this is done within the restricted capacity of the people and machines in the lab for addressing the problems at hand.

By necessity, many arguments about methodology are fundamentally controversies about the merits of machine systems. The following is from a proposal for a program of *in situ* methods for deep sea biology.

There are two basic approaches to studying the metabolism of deep-sea organisms, 1) a laboratory approach and 2) an *in situ* approach. The first method involves capturing the organisms in their environment and returning them to the surface to conduct measurements and experiments in the laboratory. Many limitations of this approach must be resolved or acknowledged.

A) The capture process, no matter how gentle, stresses the organism. This is evident from initial oxygen consumption rate increases noted immediately following entrapment.

B) Organisms captured at depth and brought to the surface unprotected will undergo decompression and temperature change. Decompression problems are being addressed with pressure retaining mechanisms that have been employed in traps and net cod ends; temperature control problems are being addressed with insulated traps and net cod ends.

C) Solar or artificial light sources can adversely affect visual pigments of deep sea organisms accustomed to low ambient light levels.

D) Organisms are generally returned to a surface ship where they experience varying degrees of abnormal motion (rolling, yawing, pitching).

E) Confinement in containers for both laboratory maintenance and metabolic measurements places physical and biological constraints on organisms.

F) Organisms are generally held in surface sea water ignoring the water quality differences between surface and deep sea. The *in situ* methodology is based on capturing and measuring the metabolism of organisms in their environment. There are limitations to this approach which must be considered.

A) The capture process, with its associated stress to the organisms, is a problem pertaining to *in situ* work as well as to the laboratory approach mentioned above.

B) When submersibles are used for *in situ* operations, the artificial lighting will have a temporary blinding effect on animals with visual pigments which normally experience attenuated light levels.

C) There is a severe limitation on the number of measurements and the complexity of the experiments which can be performed.

D) Containment of organisms in either flow-thru or closed chambers is an abnormal situation. These effects are believed to be minimal from comparisons of the observed behavior of entrapped versus free animals. After analyzing the limitations of the two methodologies, we chose to use the *in situ* approach because it allows measurements . . . to be made under conditions more closely approximating the natural environments from which the organisms have been removed.[17]

Where competing systems of data collection and analysis are not so visibly caught in exchanges over the intellectual merits of science, discussions of how to use instruments still abound. A serious rivalry developed some years ago between "Big Lab" and "Blue Lab" over the merits of tethered versus untethered vehicles for data collection. The debate was heated at the time, but now it seems an embarrassment to explain to an outsider. Today they discuss *when* tethered and nontethered systems are most useful.[18]

The stakes involved in these technical decisions may seem very small to outsiders, and perhaps in the long run for the development of science they are. But they serve vital purposes in the short run. They mark the boundaries of science by identifying improprieties in the use of instruments that could distinguish insiders from outsiders to the world of science. They also help adjust and readjust the relationship between researchers and machines in the continuing process of developing and comparing laboratory signatures.

Because laboratory signatures are so central to the research process, when researchers write proposals they pay particular attention to their proposed research methods and their relationship to earlier techniques employed in their labs. It is easy when you first look at these proposals to imagine that the attention to methods is simply an expression of the centrality of method to the logic of scientific experimentation. But the careful discussion of how the proposed work relates to earlier research done by this and related laboratories suggests the importance of signatures to research projects.

Proposals for large projects are an interesting place to look at the functioning of laboratory signatures. The expertise of different laboratories is described and presented together to a funding source as an appropriate amalgam for the proposed work. Each laboratory that is supposed to belong to the project must be justified as appropriate on the basis of past work; the laboratory's signature must be described as particularly useful for the group's efforts. Then there are lengthy discussions of how the new work fits with old work, how it can be justified as new science, not just a reiteration of old work. The laboratory's signature must be described as either growing in some fashion or newly revived by its application to a new scientific problem. The success of the project depends on the success of the signature in two ways: first in convincing funders of the promise of the project, and second in demonstrating to scientists the value of the signatory approach to problem solving. Within these proposals the micropolitical manipulation of signatures is closer to the surface, and we can see for a moment to what extent the conduct of science and its evaluation depend on the competition among signatory techniques.

THE DEROGATION OF TECHNICAL SKILLS IN SCIENTIFIC LABS

While the establishment of a set of signatory techniques is central to the competition and cooperation among scientists, the people who design and execute many of the techniques are often not chief scientists but technical support people—lab technicians or engineering groups. The work of these people makes scientific work possible, but it is socially less visible than the world of chief scientists (with rare but interesting exceptions).

The invisibility of technicians in labs derives from the social definition of scientists as individual geniuses, making discoveries with their superior intelligence, not with their lab technicians and equipment. Scientists are treated like artists, as "auteurs" (to use the film term). Like the director of a film, the chief scientist may work with a large group of assistants but has final authority over the science that is produced in the laboratory. That does not mean of course, that the scientist does the important conceptual work that makes a project interesting or develops the signatory techniques that make the lab's work distinctive. What is important is that all the work done in the laboratory is at least partially credited to the chief scientist.[19]

Socially, the work done in the lab "belongs" to the chief scientist. This system of attribution is for science (as Becker suggests it is for art) fundamentally arbitrary. As Becker notes, there is a lovely contrast between poetry readings and political speeches that points to the arbitrariness of attributions. Political speech writers are seen as providing words that "belong" to the politicians who utter them in public. In contrast, someone reading poetry is merely a conveyor of words that "belong" to the poet.[20] The arbitrariness of the attribution system is also apparent in film history. In the United States before the art film movement of the 1950s and '60s, films were thought to "belong" to their producers. This made perfect sense. Films were commercial commodities and were the inventions of the people who owned them. But after the French introduced the idea that films were art to the United States, Americans began to learn that films really belonged to the auteur-directors who were aesthetically in charge of filmmaking. This attribution system gave films a single creative force and allowed them to be analyzed using the tools of art criticism. Once people got used to the idea, it made just as much sense as the older system of attribution.[21]

In science, the "great man" tradition has made it socially important to identify a single creative force in science (or in art) You need someone to give Nobel prizes to, and someone to be heroic to high school students. So the work and thought of a variety of people, all embodying and exercising a wide range of skills, are all focused on an individual, and the work of the others is relatively obscured.[22]

The tradition of ignoring laboratory support personnel fits nicely with the view of machine tenders as deskilled. Lab assistants are easily conceptualized as robotlike followers of directives from their superiors and their machinery. But clearly they are much more. They are often very serious scientists doing much of the work that is credited to others.

Laboratory Assistants

Most of the laboratory personnel in the research groups I studied are hardly mindless technicians, pushing buttons and watching dials. Almost all have or nearly have Ph.D.'s. Some are graduate students; more are post-docs; a number have engineering degrees; but the largest group are Ph.D.'s who have not established laboratories of their own. They usually come to established labs when they finish their Ph.D.'s (sometimes as post-docs), and stay to maintain their access to advanced instruments, government funds, and the intellectual life of a successful lab. One man put his experience this way:

> Well, it was very stimulating there and the research was really great and it was fun. There was lots of money, you know, and for eight years, I was completely [on] soft money the whole time. I helped to write proposals, although I never really did. I guess I wrote one proposal on my own . . . so that part was good. But really after about two years I started looking for another oportunity. But it was hard to find one that was, that I felt was, good enough. [Interview with a junior scientist at a major university]

What laboratory assistants do in the lab can be very boring and monotonous work, but it is usually also highly skilled work. They make cultures for experiments; count crabs in pictures of the seafloor (which requires distinguishing between many similar types); develop and/or modify computer programs to fit the requirements of a particular experiment or equipment; run analyses to identify certain chemicals; build instruments for conducting a particular experiment or a particular cruise; or (in some cases) run an experiment on their own, using the resources and good name of the chief scientist.

They frequently describe themselves as translators of the chief scientists' desires and needs, making them into practical courses of action and equipment for reaching them. Very often lab technicians of this sort are given enormous authority in the lab. Outsiders may not recognize their importance and can attribute all of a lab's successes to its "head," but the technicians (and often the chief scientists) know better. In interviews, some chief scientists introduced me to people in their labs, describing them as very important men or women in their fields and saying I should interview them if I wanted to know about science and technology. Sometimes, they did this just because they were trying to pass me on to their inferiors. They wanted someone else to handle the pesky sociologist who was taking too much of their time. But other times, they touted their people because of their appreciation for what one man called "his guys."

The extent to which junior researchers run laboratories in the name of chief scientists may be indicated by a bitter comment from one chief scientist I interviewed. When I asked him what it would be like for him if he lost all his funds, he said it would be an opportunity to get back to doing some science. It was a bad day for him, when he said this, because he was facing a massive cut in funds, but the sour remark is still revealing. It suggests the extent to which chief scientists rather than laboratory personnel may be deskilled by the division of labor in science. A large laboratory establishment can make administrators out of chief scientists.

Chief scientists are often particularly dependent on younger colleagues in their lab to keep them up to date technologically. The junior people who fulfill this function are often so important to the lab that if they leave, they are sorely missed.

[After I took a job of my own] amazingly, you know, I still had access to the machine, to the lab and the mass spectrometers and stuff at [Big Lab], so I could go down there and still analyze samples. . . . I think that [arrangement] was sort of mutually beneficial because [my former boss at Big Lab] was, I mean I was, the guy that kept all the equipment running and everything else. So, it was fine if I wanted to come back because if they had some problems they could ask me about it, so. . . . I mean one of the things I'm good at is building and operating complicated equipment and getting it to work right, and stuff like that. I mean not every earth scientist is, takes an interest in that or is as good at doing it. [Interview with a junior scientist at a major university]

It was not sentimental feeling that tied this former technician to his old boss. It was his importance to the laboratory. The fact that he was able to get his own position testifies to his abilities. After all, most young scientists are not able to secure their own positions; they become permanent laboratory technicians instead. This does not mean, however, that junior researchers who become permanent parts of their labs are any less valuable to their labs. They get permanent positions in them *because* they become essential to the functioning of their workplaces. The best of them are described as geniuses who do not have the political or interpersonal skills to manage laboratories, but are able to solve problems better than anyone else. They are no less central to the lives of their labs than their chief scientists, but must work in conjunction with a more entrepreneurial scientist to be able to survive in the soft-money funding system.

Over time, these people may become increasingly identified as extensions of their Principal Investigators, just as their equipment is treated as an extension of their senses and manual abilities. In spite of their centrality to the research process and even with the inclusion of their names on journal articles written with the chief scientists, they end up as the "et al." in citations. Their gifts are absorbed into the laboratory signature that is written in the "hand" of the chief scientist.

These career patterns, coupled with the routine, manual character of much of researchers' jobs, makes it easy to see these people as the "working class" of laboratories. Most of the value of what they do is socially assigned to the chief scientist, and what they do with machinery is devalued, while chief scientists are congratulated for their technological sophistication, if their technical support is good, and are seen as first-rate scientists, if they have been able to assemble a talented group of co-workers.

Research Engineering

Young researchers and other research technicians in laboratories suffer from lower status for working at the boundary of science and technology, using manual as well as intellectual skills. So too do the ocean engineers who assist in laboratories or work in an engineering group in a large oceanographic institution. Vannevar Bush reinvigorated the nineteenth-century elitist vision of science after the Second World War, and in doing so he reinforced the view of technological innovation as subservient to and dependent on scientific innovation. In Bush's proposal, he argued that technological innovators alone could not be counted on to solve the problems of the state. They needed scientists to teach them about basic characteristics of nature that could affect both technological problem solving and equipment design. He placed engineers on a stratum below scientists, arguing that the well-being of the nation depended on science more than on engineering.[23]

The subordination of technical innovators to scientists not only was important in legitimating the development of an elite reserve labor force of scientists after World War II, it remains important to the organization of scientific research today. In most laboratories today, ocean engineers are subordinate to a chief scientist (or an engineering group is subordinate to a whole gamut of chief scientists in a large research establishment with a separate "shop" or set of "shops").

This hierarchical distinction, because it was used by Bush to justify funds for science, obviously is important to the scientific community for practical as well as symbolic reasons. It helps confirm the elite status of scientists and gives them the greater authority and fiscal advantages.[24] And more than this, the hierarchy has become taken-for-granted and seems somehow almost natural. Many scientists I interviewed genuinely believed that technological innovation was not particularly valuable as an activity in its own right. Ironically, we find this view expressed by a scientist known for his cleverness in devising new research equipment and techniques for deep ocean biology:

> I have very little use for people that create new machinery and then look for a use for it. I mean, the only time I've ever made a new piece of machinery is because I had to have something. I mean it was the combination of seeing some kind of real interesting problem and seeing a way to do it. But there's certainly a lot of people in the field who will develop very fancy pieces of machinery and become very committed to that piece of machinery. And that's not ever been my approach really. If anything, I decided I really don't particularly like making new [equipment]. I mean, I do enjoy doing it, but if at all possible, I'd rather buy it off the shelf. Because it's just a huge amount of work and effort and it's a lot of just creative energy, of just sitting there and just picturing, you know, how's this thing gonna work? And then, translating that into the hard things. [Interview with a senior scientist at a major university]

The man is obviously trying to distance himself from engineers, while also admitting that he develops his research with technological as well as conceptual innovations. In the course of my interview, he presented me with lovingly detailed descriptions of the machines he and his staff put together to do their work. He was clearly proud of the ingenuity used in these efforts. He just wanted to make it clear that this cleverness was not meant to be the valuable aspect of his research. He saw it as motivated and justified entirely by its service to science. He states quite explicitly the Bush view of the unequal relationship between science and technology.

Most labs reproduce in their organizational structure the subordination of technological innovation to science. Usually the engineers who work in these labs are subordinate to their chief scientists, and they have little chance for a separate career in science. They are anything but favored by the funding system. Their jobs, much more than the jobs of their Principal Investigators, depend on the year-to-year continuation of grants. The Principal Investigator may be able to support himself or herself by teaching or doing less technologically complex research, but the technical support personnel who are hired to tinker with and use expensive research instruments cannot (for the most part) get independent funds, and thus they are only employed as long as chief scientists can afford them and want to keep them.

Because of this job instability, many of those doing engineering work at scientific institutions are young. They are there because they cannot get their own funds, and they do not relish the idea of working as a cog in the wheel in some big engineering group in business or in the military. They prefer the flexibility of working in a lab where they may have a project of their own to do with the Principal Investigator. Some of them are new Ph.D.s who are being what sociologists call "cooled out by the system." They are being removed from competition for jobs as primary researchers. More of them are engineers who have just finished school, or engineers seeking refuge from business on a permanent or interim basis. A small number of them are seasoned science engineers. They usually have been involved in very large labs that juggle large numbers of projects, so they have always found the money to keep themselves going.

As they get older or the labs they have been involved with dissolve, both seasoned engineers and former graduate students often look for something else to do. One favorite alternative career is to start a company to supply ocean equipment to whoever will buy it.

So at [the same time that I'm writing my dissertation and doing the image analysis work] I'm starting my own company now in the outside world. I started it about a year and a half ago. So just for survival I've been consulting and designing and building things for the recreational diving industry. Underwater flashlights, and I have a patent on a weight belt design which was licensed to a company in [inaudible]. So I'm very heavily involved in the technology. We just got a contract to build a ten thousand dollar deep sea camera system for [a local principal investigator] which he's supposed to get. And it's something that I wanted to do for a long time. . . . See, there aren't any economical systems. For a few vehicles you'd like to have ten or twenty of these, and you go out and just drop them off all over the place from the ship. But at twenty or thirty thousand dollars a pop (and they weigh 250 pounds each), it's totally not feasible. You're not gonna be able to build them for fifty dollars each, but I think we'd be able to build them for maybe thirty-five hundred dollars each. Which is like about a factor of three or a factor of four difference. [Interview with an ocean engineer and graduate student in oceanography]

During the 1960s, even a small industry developed for companies manufacturing research equipment for ocean scientists, and many engineers involved in ocean research began private firms.

Then in 1962 I decided to strike out on my own and moved to Cape Cod and started a company in the barn in the back of my house. So the reason I moved to the Cape was because it was near [Blue Lab] which is an oceanographic institution. And I already knew a number of people down there, so I started doing business with them. Then I didn't get into the underwater camera business again for about seven or eight years, and I'll come to that. But I was manufacturing instruments for at first just [Blue Lab]. The way that developed was very interesting because I would go down to the labs there and talk to the scientists and found that there were instruments that they had developed for themselves, just building one at a time. And when I found there was an instrument that they had to keep building more and more of, they would ask me if I would build them for them. So I would do that, and then I would ask them, well, who else uses the instruments? And then they would give me a whole list of their peers in various parts of the world. A good example is a thing called a time depth recorder. This is a little instrument that's clock driven and it's [sensitive to] pressure in the ocean. And of course pressure is related to depth, so if you put this instrument on a plankton net that you would be towing through the sea, then when you get the net back you know how deep the net has been and how long it was there. So when we inherited that instrument it was an incredibly crude thing. . . . It had a clock which is made out of a watchman's time clock. Really a lot of homemade stuff. So when I took that over, we used the stainless steel housing and professionally made a pressure element and a very big clock and so on. So we made thousands of these, now selling for about a thousand dollars a piece. So just that one instrument we sold over a million dollars worth of, and the fun of it was that it was a little item out of a laboratory at [Blue Lab]. And then I did the same thing with chlorine equipment for sampling and setting up on the ocean floor and finding that they had developed their own chlorine equipment and really didn't enjoy reproducing it. They needed somebody to supply to them, so we did the same thing. [Interview with an ocean engineer and entrepreneur]

Since the late 1960s these companies have done poorly because of cuts in government funds for researchers that have limited what scientists could spend on equipment. As a result, the major science engineering firms have turned to government or the oil industry as the new, central markets for their machinery. A few smaller firms have, alternatively, tried to develop products that would interest amateur divers. In either case, supplying equipment primarily to researchers is a less attractive way of making a living than it was twenty years ago. Still, developing equipment remains an option that some ocean engineers try when other sources of funds seem too insecure or frustrating.

A few science engineers without science degrees who work in oceanographic institutions are able to overcome prejudices against them and land their own contracts, but these are rare. Since NSF has placed more emphasis on engineering research, they are not as rare as they have been, but they still remain an embattled minority among oceanographic researchers. It is also instructive to realize that, while this elite group of technicians can gain some independence over their research and development activities, they generally still see themselves as serving the technical needs of scientists. While they may be structurally threatening the system of scientific dominance because of their independence, by claiming they are just there to serve, they reproduce the traditional hierarchy; hence the Bush formula (and the prejudices that formed it) are maintained.

Engineering Scientists

In spite of the prejudice against technological "tinkerers" and the organization subordination of technicians in labs, there are still some researchers who build major laboratories around technological innovation. And they provide the most clear examples of the ambivalence toward technology in science. In oceanography, these are the people who devote themselves almost exclusively to designing and providing the scientific community with ocean instrumentation for research. They make machines to facilitate or reproduce laboratory operations at sea, or they create the vehicles and sampling systems to make data collection easier in the ocean.

In spite of their usefulness to the research community, the people who run these labs are often treated by their colleagues as somehow less serious than "real researchers." When I asked one informant whether or not another had gotten tenure yet, he said significantly that he had, but only in engineering, not in a science department. In spite of prejudices against them, many researchers find this kind of career very rewarding because there is actually a great deal of money around for technological innovation by scientists, as long as they have science Ph. D.'s (or a science publication record that equals it) and can thus formally embody the subordination of technology to science.

The wealth and low status of these innovators indicate their ambiguous position in the world of science. They do not take the pursuit of pure knowledge and success in scientific publication as their major goals. They may be *in* science, but in some fundamental way, there are not *of* science. They sit on the boundary of that world, giving that world needed intercourse with the world of technological innovation and state-of-the-art engineering. Their marginality to the culture of science is signified by their low status in the world of science; their importance to the practice of science is signified by the amount of funding they can secure.

Getting funds to do technological innovation is quite unlike getting funds to do regular research, so the laboratories that devote themselves to this kind of work are quite different from many others. The most successful of them tend to be relatively large, employing up to one hundred engineers and technical support personnel to make and move machinery. Young scientists who are capable of doing this kind of work and who find funds for it are particularly lucky (according to their peers as well as themselves) because they face a more stable financial future. If they can design machines that will allow scientists to do new kinds of research or to make necessarily fundamental measurements more efficiently, the government has interests in supporting them, particularly if the measurements have more practical uses, too. And if they make enough inventions that are used by others, they can profit from rentals and from patents. This gives them the opportunity to generate funds outside of the government funding system, and thus provides for them more autonomy in their work.

In addition, technological innovators have opportunities to seek applied funds that are not applicable to most research. This sometimes gives them greater choice about where to seek funds, and can give them the ability to put together different pots of money to support complex and expensive projects. As a result, their relationship with the soft-money funding system may be less burdensome than the one experienced by their peers who do more traditional research projects and less engineering, creating perhaps enough jealousy in conventional researchers to motivate them to deride engineers, and ratify the traditional hierarchical relationship between science and engineering.

Without this motive, it is a little hard to understand the depth of the antipathy toward engineering found among many scientists. After all, the engineering groups I looked at seemed to be doing a service for science. They seemed to try to apply new developments in technology to the research problems of scientists, seeing what new machinery could be modified or built from available parts. They have often produced exactly the kinds of equipment for research that scientists needed but could not themselves get from the military or business. These machines then became resources that scientists used to plan their own projects. The following account by one scientist of how to do proper geological research along spreading centers shows how he uses different kinds of research equipment as *conceptual categories* for thinking about research:

> The sequencing would be to begin with this multi-beam, sea beam type sonar, and survey the area out in very careful detail. And depending on what you want to know, you can do other geophysics like magnetic measurements. The next step then would be to form the maps that you make, like the areas that need to be photographed with a towed photography system [like Angus]. It's a very foot system compared to, say, the Deep Tow. After you have done that, then you can select areas where you may want to know more by means of the sophisticated Deep Tow which can tell you how much sediment there is in an area that looks just like a bed of sediment. It may just be just a thin veneer of sediment. The Deep Tow would be able to tell you that. And then you get deeper. And with a Deep Tow survey, you want to do most of your sampling of rocks first by surface techniques like dredging or coring until you learn, again, to get an idea of what is there. And then finally when all that is done, you can sit down and say, now these are unsolved questions that I really need to see and manipulate, to go down for. And one of them is sampling hot springs because they're simply too small, you see. So [then you go in with *Alvin*]. [Interview with a senior scientist at a major university]

If you talk to the people in labs that devise or customize research technologies, they will tell you that science is limited by technological factors only because they are limited in funds. They say they can make the machines needed to answer many of the research questions scientists would like to address, but they do not have the money. They repeatedly emphasize the dependence of scientists on technology and techniques, rather than the reverse formula dictated by tradition and the Bush report. They are fully confident of the importance of technological innovation to the scientific enterprise, and find no reason to apologize for what they do.

Recently, members of just such a laboratory discovered the *Titanic* on the floor of the Atlantic ocean. In this case and others, these researchers are able to make themselves and their work quite visible, even when it is not really "science." The public appeal of exploration using the latest kinds of equipment is great enough that these people probably gain greater prestige than their more "successful" colleagues in scientific disciplines. And they have lots of evidence to show how the work of more traditional researchers is improved by access to their equipment. They also have all the income generated from rentals, and popularizations of their work.

While the ideological and organizational subordination of machines to science is clear, the symbiosis between scientific findings and machines is close enough that it is sometimes difficult to say which is inspiring what.

> The main difficulty that we had, and this is technical, is that the echo sounders we had available at the time sent sound down in a cone with a top angle of thirty degrees, and so you see by the time we went down to two thousand or three thousand meters, which is where the crests of the midocean ridges are, you were shining sound on a very large area all at once, and then it comes back and you don't know whether you've got this point over there or over there. What it basically means is that anything that is much smaller than a couple of meters across, you can't be sure that we can see it. And so the amount of detail that you can get in that is very severely limited. And as a result at [Big Lab] and at [Blue Lab], people started developing equipment to deep tow along the bottom, that would be maybe ten to fifty feet off the bottom, and then you wouldn't have that problem. And you could get photographs that would see something and you could do echo sounding and observe, and you can get magnetic measurements and all new equipment. [Interview with a senior scientist at a major research institution]

When researchers resort to developing equipment to explore new areas of the ocean, they underscore the dependence of scientists on the instrumentation at their disposal. But when scientists cannot make the machines they need and instead develop projects to make use of available equipment, they illustrate most clearly the extent to which scientists are technological "dependents," shaping their research to fit agendas written by engineers in other fields. The character of this dependency is difficult to understand at first because the technical side of research so often is left out of conversations about science. Scientists still act as though engineering simply serves science and has no separate life. They believe this even while letting the characteristics of the machinery available to them shape the trajectories of their research projects. They learn to think *through* their machinery, to use it as a guide for their scientific imaginations, so it is easy for them to experience it as transparent. They do not notice the people who make the machines, those who use them, and how their laboratory signatures are shaped by technological constraints.

CHAPTER EIGHT

Expanding the Domain of Science

IN DOING RESEARCH scientists may be in some sense engaged in a training exercise for improving their value as a labor force, but they are above all (and in their own eyes) doing science. They address themselves to the social world of science to dominate some portion of that world, while simultaneously promoting science itself. Research may embed the practice of science into the structure of funding (because of its machinery and cost), but it also contributes to a transformation of nature into data (data being representations of the natural world that accommodate scientific analysis), and the latter are a source of power for science. Scientists use research to enlarge the domain of science, applying scientific principles to the study of areas of nature neglected before (or left in the hands of folk or popular culture), using their results as new cultural capital for science, and applying it to the internecine struggles at the center of the social world of science.[1]

Through the transformation of specimens with scientific methods, bits of nature are translated into elements of culture. In the world of culture, specimens become social instruments in themselves. They are hoarded and brought out by scientists to further science or their own careers. They are markers of domination, demonstrating that one has the capacity to control them. Like the Oriental ornaments collected in European homes in the eighteenth and nineteenth centuries, like the exotic flowers nurtured in the gardens of the same period, or even like the silks and spices valued earlier, rubbery grey specimens floating in laboratory jars are to trained eyes exotic marvels from distant worlds; they are emblematic of the same cultural reach and ambition.[2]

Just as consumer goods from distant lands were measures of the power of the West to extract materials from other cultures, laboratory specimens are measures of the power of science. This is true both on the symbolic and practical level. Bringing back rocks from Mars is a technological feat with its own symbolic contribution to our understanding of scientific prowess. To the extent that scientists use such evidence to make theories about the origin of the solar system or the earth, they add to the power of science. In both the extraction of specimens and their analysis, scientists alienate nature to serve the interests of science. In doing so, they make the institution of science less vulnerable to outside manipulation by increasing the cultural capitalization of their expertise. As long as scientists know nature (in their peculiar way) better than others, and as long as their peculiar way of knowing is socially valued, the fiscal dependence of scientists on the state is partially offset.[3]

To acquire intellectual property for science, researchers do not just randomly collect samples and contextualize/reconceptualize objects in the natural world. They interact with nature using signatory techniques, which orient researchers to and limit the responsibility of scientists for explaining the unknown. With their techniques, scientists learn how to make models of the objects of their research. These models become the property of scientists and the markers of the domain of science.

While people usually distinguish between descriptive and theoretical models in science, I think it is more useful to talk of *ostensive* models to destroy the false dichotomy between describing and theorizing. Ostensive models characteristically do both. They are referential, like descriptive models, attempting to represent some perceived object, be it a particular laboratory specimen or something more grand and abstract like water circulation in the ocean. At the same time, ostensive models explicitly attempt to organize perception (and hence understanding) of the object in the way they describe it. They express a point of view for examining the object as well as the object perceived from that vantage point. Hence the term "ostensive model" expresses more precisely the interactive nature of model building—the role of the observer in defining the object. The term "ostensive," then, does not imply (as "descriptive" does) an attempt at producing realistic fidelity between model and object. Certainly ostensive models must be judged in part on their capacities to describe an object (because if they go too far afield from the object, they have no critical value). But ostensive models are primarily evaluated on their usefulness in *reconstituting* the object from a particular point of view so that its important features stand out.

I borrow the term from the art historian Michael Baxandall, who uses it to describe analyses of artworks.[4] Critical commentaries on paintings are ostensive models. They necessarily refer to the work, and they are pieces of critical analysis assessed in comparison to the object of study. But they are not meant to be simply descriptive. They attempt to highlight salient characteristics of an object to make sense of it. They are "about" the object to the extent that they try to make the importance of the object in some scheme more palpable to the reader. At the same time, these models are clearly also "about" art history; they present the object in a cultural tradition in order to further that tradition.

Models in science are much the same. They refer to *categories* of objects and *explanatory schemes* as well as to the objects themselves. They are abstracted from the natural world through interaction with the culture of science and its theoretical traditions.

So what happens, in this formulation, to the distinction between theoretical and descriptive models, used so widely by scientists? We can still talk about degrees of abstraction from or alienation of the natural world in model building, capturing the distinction between models designed to answer fundamental issues in the fields of science (so-called theoretical models), and those attempting to simplify complex observations and make them available in model form for further analysis (so-called descriptive models). By thinking of both of these efforts as ostensive model building, we no longer think of one type of work as apples and the other as oranges. Instead we see them both as designed to enrich science through a reappropriation of nature. They make "nature" part of the world of scientific culture, part of the cultural capital that researchers can use to enhance their social standing.[5]

There are different levels to the alienation of nature for model building. The simplest one is a "framing" process:[6] noticing something in the natural world and giving it a place in the categories of science. Framing may not entail a physical decontextualization of objects, but it does require a cultural one. When nature is alienated, the "meaning" of objects derives not from their location in nature but from their location in scientific schemata; they are examples of a scientific category or an anomaly that requires explanation in the language of science.

There are also many forms of data collection that entail some physical as well as cultural decontextualization of objects. These represent another type of alienation/reappropriation of nature. Animals, rocks, water samples, or whatever are brought into the labs as representatives of the natural world. In the process, they are decontextualized; they become parts of laboratories. This is most obvious when specimens are strewn about laboratories as emblems as well as working parts of the labs. But the same holds for *in situ* measurements in which the laboratory is recreated (using complex instruments) in the field. What is important is the recontextualization of samples within meaning or measuring systems; this changes their status. They become part of the social rather than the natural world.[7]

The alienation can proceed further. There is a restructuring of nature to make it amenable to scientific analysis. A photograph is a restructured version of the natural world. It is obviously not nature, not even a piece of it; it is a transformed version of it that is more portable, measurable, and hence interpretable than the original. Other transformed kinds of data are tagged fish, chemically altered objects (like frozen animals or ones preserved in solution), stained pieces of tissue, or symbolic representations of rock formations placed on a map. These transformations yield models of some natural phenomenon, rather than the original. The models may be physical or conceptual; altered objects may come to represent and comment on salient features of some larger phenomenon just as conceptual models do. In either case, the model stands for nature within the world of scientific culture.[8]

This brings us to another level of model building, an alienation in a more technical sense: the use of nature (now data) to oppose itself. Nature is made to "give up her secrets" (as the cliché goes and one informant told me). This metaphor expresses an antagonism toward nature that fits the views of many feminists, environmentalists, and animal rights activists about science. In the world of scientists, this level of model building is the use of nature to empower scientists and science (and perhaps, directly or indirectly, business and the state). When data are used for revising scientific theories, for directing efforts at resource extraction (like mining manganese nodules), or for simply expanding the territory understood through the language of science, this is the highest level of alienation. The exploitation of nature in these cases may or may not have immediate effects on the natural world. What makes this stage of alienation different from the last is that at this level scientists extract *powers* from the natural world by learning to control them intellectually. They put their images of nature to work to serve their science or its applications.

In a severe mood, one can argue that scientists abandon the study of nature for the study of their own creations, once they make models.[9] They begin secondary and tertiary analyses from the models rather than from natural forms. So in a technical sense they end up studying the artifacts of their own technology rather than "nature." Of course, this does not mean that model building is always or even usually deceptive, misleading scientists and those who listen to them. It may be full of human fancies, but this does not make it necessarily a folly. The literature on the processing of visual perceptions suggests that this kind of cognition does not so much distort perceptions and delude people as give them something to perceive. Human beings do not "perceive" without making models and using them to interpret what they see. Models, then, allow scientists to think about natural processes, learning from them how to alienate nature and make it serve the interests of scientists.[10]

While ostensive modeling is the means by which scientists alienate nature to serve science, the models they produce may be used by others. If researchers reach only the second or third level of alienation, the models they produce may be easily "read" and used by nonscientists. If scientists find new seamounts in the Pacific and describe them on maps of the seafloor, they may well intrigue the Navy more than other scientists. On the other hand, all models are selective in defining salient characteristics of the natural world, and scientists usually select them according to values emanating from the world of science, so in the end, most serve science more than the state. An ostensive model of how the blood of fish handles toxic chemicals from seawater may draw attention from few people in government concerned with toxic waste, but for the most part the research is simply too specialized to interest most nonscientists.

There is one exception to this rule, and it is an unusual one. Some of the most rudimentary mapping of deep sea terrain, basic exploration, is probably more interesting to the general public than to the state or science. It is not so absorbed in the world of science as to be inaccessible to outsiders, and it can easily be contextualized in a tradition of romantic storytelling, placed in the popular realm, and not just the public one. For many television viewers, Jacques Cousteau made the act of exploration very compelling in just this way. Other scientists want to do the same thing, although they are exceptions to the rule.

I think of myself as an explorer and it sounds a little corny, I guess, but the concept of exploration is an exciting one, that's it. It always had been a driving force in, you know, society for a long time. And most of the public think it's sort of over with, you know, that exploration was somehow all done a long time ago and that all exploration today is done by astronauts in space. And I don't let that bother me, but they can have their opinion, but I look at our planet as a fairly unexplored thing. Some of the statistics that you hear is three-fourths of the planet is underwater. The average depth of the ocean is twelve thousand feet. And that's the *average* depth. . . . What happens is that you take a map like the one you have on the wall here and it gives you the impression that it's been mapped because there's a map. But Sir Francis Bacon had a map too. Pretty bad. And particularly in the southern hemisphere, there's very little known. . . . Now see all these strings of volcanoes [he points on a map of the seafloor to a string of volcanic seamounts]. That's because the ship went there. But there's none [no volcanoes on the map] there [he points to stretches of empty seafloor]. It's not that there isn't a volcano there; a ship's never been there [to map them]. So our sounding density in the southern hemisphere is less than one percent of the [area]. . . . People sort of don't believe it when you say that we know very little. They say, well, you're [just] trying to get money. And we hear all these stories that there's this missile threat; there's the hunger threat; and there's this threat; and now there's a we-don't-know threat in the ocean, who cares? There's a classic cartoon that was in the *New Yorker* where it shows four women on the sofa having tea and one said, "Mildred, I'm sorry, I just don't care about the deep sea." Say okay, all right, don't care. What motivates me is not to make Mildred happy. What motivates me is almost the romance of exploration, to know that when you're down in a submarine and you're looking out of the window, that you're the first human to ever see that. And that you don't know what you're going to come into your view of the windows. That's sort of an eerie, a natural high, a natural excitement to know that you're the only human being . . . it's the first time that fish has ever had light on it, 'cause everything down there's in total darkness. So when you flash a light, the headlights of a submarine, it's the first time that light's ever hit that subject, which is sort of strange. [Interview with a senior scientist at a major oceanographic institution.]

There is excitement, personal glory, and even money to be made by explorers like Jacques Costeau and his counterparts in this country. Looking for buried treasure or finding the *Titanic* can be interesting and well-publicized work. In the age of mass media, this is the science that primarily enters the public realm, and it has a large enough following to be quite attractive. But it is not the science that gets the most attention in the social world of scientists because it has limited use in expanding the domain of science. Finding an interesting corner of the world to study is not enough to enrich science. Scientific definitions of these new realms must be established if they are to be part of the domain of science, but exploration is a first step. By defining what there is in the world that needs scientific analysis, this kind of exploration acts as the base for scientific model building.

THE DISCOVERY AND EXPLORATION OF HYDROTHERMAL VENTS

Deep ocean researchers of the postwar era are particularly interesting to study if you want to see how scientists, using soft-money, have mobilized their capacities to model nature to expand the domain of science both physically and conceptually. They have been extending their grasp of the ocean in very dramatic and clear-cut fashions. They have found mountain ranges, new species of animals, and chemical processes that no one ever knew existed, much less tried to interpret. And they have been very successful in recent years in developing more sophisticated theories to explain the geology, chemistry, and to some extent the biology of the deep sea. All of these efforts have had some value to the state, but they have retained special significance for science. To the extent that scientists have trained themselves to use Navy submersibles, they have made themselves a more valuable part of the scientific labor force; but to the extent that they have contributed to the elaboration of plate tectonics, they have empowered science more than the state.[11]

Scientists have done this (as usual) by making models of natural features that were either unknown or poorly modeled before. More than most parts of the world, the ocean has been relatively understudied. The recent breakthroughs in ocean research have been able to produce (at least in many cases) far-reaching results because they have simply replaced complete ignorance (or bad theory) with some empirical analyses. The shape of the mountain ranges running through the major oceans of the world in itself has been a relatively recent discovery. Clearly, the level of ignorance has been great enough to allow rather great empirical and conceptual leaps.[12]

Included in the list of major "findings" has to be the discovery of the hydrothermal vents on the seafloor. While geologists had suspected for some time that hot water ought to be seeping out of the rocks along the mountain ridges in midocean and went to find them to confirm their theories, they discovered something much more dramatic: dense animal populations growing in streams of hot water in which were dissolved large quantities of toxic chemicals. The fascinating puzzles and opportunities presented by the vents made them important research sites for oceanographers.[13]

When scientists went down to study the hydrothermal vents, they drove to the seafloor on the deep submergence vehicle, *Alvin*. Videotapes made automatically on these dives recorded much of the conversation and many of the activities of the scientists. They constitute a rich source of data on this particular piece of territorial expansion for science. The stationery camera on the front of the submarine did not always record the movements of the vehicle's robot arms, particularly when the scientists were setting up experiments on the ground close to the *Alvin*. In those cases, the activities went below the camera's line of sight. But the movement of the *Alvin* itself is well documented along with most of the conversation and many of the activities of the scientists.[14]

The vents studied during these dives are essentially hot springs on the seafloor. What surprised scientists first about the vents when they discovered them was the abundance of life around them—clams, mussels, and large worms. Nothing in the literature predicted such a thing; there appeared to be too little food in the deep to support this amount and variety of life. What were the animals living on? How did they get there?[15]

In addition, researchers found that the water escaping from these vents carried high concentrations of substances dissolved from the rocks under the ocean. Some circulation system was churning hot water through benthic rocks and bringing minerals into the seawater in some quantity. How did the character of the seawater emerging from these vents relate to the metal deposits on the deep seafloor? And would this water finally explain the chemical composition of seawater?[16]

To begin to answer these questions (and successfully model these sites), scientists had to think about how to manage the practical and cognitive problems of doing work at these remote and perplexing locations.

How could they get there? What could they do once they got there? What were the hazards of heat and chemicals in the water? What equipment could work in the inhospitable environment? And how could they put together effective experiments?

They also faced some conceptual modeling problems. What was a vent field? Did vents always have hot water coming from them, or could they sustain life just with cold water expelled from the rocks? What about "dead" vents, where there were lots of clam shells but no living ones? How did live vents come and go? What were the salient features of vents? Was heat more important than chemicals spewing from the vents, or vice versa? And what constituted an important discovery there? These and more questions needed some answers before the researchers could comprehend these sites. And only if they could answer them could they begin to make this kind of natural phenomenon part of the domain of science, a source of its empowerment as a body of knowledge, and a territory known to the world in scientific terms.

THE RESEARCH PROCESS

Studying dive tapes from the *Alvin* reveals that scientists use signatory techniques to make their models of nature. Signatures are the primary pieces of culture with which researchers interact with the natural world. Scientists come equipped with a set of instruments as well as theoretical preoccupations from their laboratories that they take with them when they go down to the vent sites.

Symbolically, scientists make claims to areas of the seafloor with their machines while they make conceptual claims with their own peculiar research designs. Scientists on *Alvin* routinely put down markers and transponders to tag research areas and make precise navigation to those sites easier. The vent fields are changed by the presence of these machines. They begin to become less natural environments than areas that have characteristics of both the natural world and the social world of science. The expression of signatory techniques through the manipulation of equipment gives scientists a way to assert their culture, and not become overwhelmed by the scale of the ocean around them. By involving themselves in the use of their machinery, they protect themselves from the assault of sense impressions at the sites. They also become carriers of their culture (in traditional missionary fashion) into the far corners of the planet, and they deposit bits and pieces of it around them in a reassuring way.

Most importantly, they use equipment to structure their encounters with the ocean in an interrogatory form. They pose questions with their

machines (based on their scientific training) about nature. (Is temperature or water chemistry more important to the vent's ecology? Measuring
one of the other is a signatory "bet" about the answer, and the equipment taken by a laboratory down to the vents will express the relevant
question for that group.) So when they put down a simple piece of equipment, they are in fact setting up a much more complex interaction, which
is the basis for model building.

The following excerpt from a dive tape made on the *Alvin* illustrates
the marking process by which researchers claim an area for an experiment. It shows how disoriented and seemingly random the procedure
can be, but how it also gives purpose to the dive.

#2: I haven't seen a good place where water comes out yet. If you see any we
can drop the pump down there

#3: Okay, look to those clams right over here. You've gotta look real hard and
you can see real shimmering water. See?

#2: Uh huh.

#3: I'm not sure you'll see any more than that.

#2: Okay, I'm all in agreement to use an area like this one.

#1: To put the pump out.

#3: Okay, this is the best area that I know of. I mean, this is the area that we
were all talking about coming to. This is the . . .

#2: Okay. Let's use that big hole right next to it. [The nearer one] is too dan-
gerous to use. . . . It could be great trouble for us [inaudible] in the hole. But
that's . . . that [other one] would be fine with me if we . . . put the pump out
here and . . . and leave it for an hour.

#1: Okay, do you want . . .? Do you need a temperature reading . . . when I
set the probe or . . .?

#2: Yeah, it would be nice.

#1: I know it would be nice, but . . .

#2: If it can't be done, it can't be done, you know.

#3: I'm gonna move over here. I think I see more. More clams, and a little
more hot water.

#2: That one's good.

#3: Now you may, pretty soon, be seeing the white smoker.

#2: Is it that . . . close?

#1: There's a clump of tube worms and it looks like, just on the other side of
the tube worms, is something . . . artificial kicking up . . . it might be [inau-
dible]. I don't know where . . .

#2: Okay, here I see warm water almost everywhere.

#1: Here's some of the clams on the mantel stuck *way* out. [PAUSE]

#3: You see that pit there, Arlen, the hole? There are a couple of tube worms and a lot of warm water coming up out of it.

#2: Okay.

#3: Is that sufficient for me to just throw the nozzle in there?

#2: Yes, yes. That's fine.

#3: There's nothing to stick it in around here.

#2: That is fine.

#3: Okay, I think we can try to get a temperature when we come back over here, if we have time.

#2: Yeah, right. I can see now . . . coming out of the clams, like a bag, hanging out.

#1: Yeah. Yeah. That's some [inaudible] coming in.

#2: Right. Yeah. I can see it here, looks right too.

#3: Okay, so I'll just throw the thing down that hole.

#2: That would be fine.

The scientists recorded here move along the vent site, not looking around randomly to learn what they can about the hot springs, but looking *for* a place to do a particular experiment: to collect water samples for analysis. When they "find" an appropriate site, they identify it using on the one hand, scientific theories of what is important about the vents, and, on the other hand, models of what the vent areas are like. They use these conceptual schemes to shape their perceptions of and interaction with the natural world. They do not see a strange world of wonder and awe out of their windows; they see a place to put down a pump to see how effective a signature can be in answering fundamental scientific questions about the vents. They apply their signatory techniques to yield new ostensive models of the vents for explaining the characteristics of hydrothermal processes and the life forms they support.[17]

There are a number of other examples from other kinds of deep ocean work in which we can see scientists interacting with the natural world in structured ways (using signatory techniques) to try to address models in their fields. One woman used "wood islands" to determine what wood-boring animals existed on the seafloor, and to test theories about how quickly they could destroy a known quantity of wood. First, she put onto the abysmal floor the "islands" made of timbers strapped together. Then the borers were left to do their work. Later, the islands were retrieved, and the changes in their weight and density were taken as measures of the borers' activities.

It is easy to think that the borers who ate into these islands were doing something perfectly natural, but what they did depended on the artificial introduction of great hunks of wood. The results were partially natural and partially artificial. Boring was natural enough for these creatures, but boring into these islands was not ordinarily an option for them. Getting them to do it made the animals participants in a modeling process intended to imitate their "normal" behavior in a more controlled fashion. One could argue that, in this kind of experiment, scientists get organisms to act abnormally so that they make good data; they are coaxed into entering the culture of science for a while, where their behavior can be measured in scientific terms.[18]

Another signatory type of research used for vent work was a photographic survey. A laboratory of marine biologists did surveys of animals near the vent sites to study the composition and distribution of this population. What scientists brought back from the seafloor was not animals, but rather pictures of animals. The whole population could not be brought to the surface; only a model based on photographs like this could be. So employing these photographic techniques, scientists made an analysis of animal populations possible. In doing the work, the researchers faced some difficulties. They had trouble, for example, distinguishing between related species of crabs from the photographs because, when the images were magnified, the grain obscured relevant details. On some occasions, the animals would climb on top of one another, and a picture taken from overhead could not record their presence. Obviously the model of nature drawn from these pictures had some problems as a result of the limitations in using photographs, but the surveys still provided a model of animal distribution that could then be compared to models of temperature variations and maps of chemical changes in the water to test theories about how the animals derived nutrients from the environment.

Other systems measured the vents in other ways. Animals were caught in cages where their metabolic rates could be studied in one fashion or another. These "readings" rather than animals themselves were then brought from the deep. Video images of geological formations were recorded during cruises, with images being coded according to a "geological video atlas" so that their labels as well as their images would be available for analysis. Bits of animals, water samples, and chips from rocks were subjected to chemical analysis. In all these ways, signatory techniques were applied to the vents, yielding ostensive models of them and addressing theories about their formation and ecosystems.

Very often the scientists used a series of modeling processes to try to arrive at a scientific model they deemed worthy of publication. They would begin with the simplest attempts at trying to make sense of their environment, perhaps gathering samples and working on them, subsequently using the resulting model to organize an experiment (i.e., decide where and how to put equipment), and finally using the experimental findings to address theory. The stages would not be used as cleanly and sequentially as this, but they are useful in dissecting the types of modeling that go on in the research process.

First-Stage Model Building

All kinds of research begin with rudimentary forms of first-stage modeling, naming, and framing of perceptions before scientists can work up to more theoretical modeling. Clumps of clams must be identified and counted before they can be accounted for analytically by complex extensive models. They must be reduced to density measures before they can be compared mathematically with the heat of vent water or the chemical composition of the seawater at a vent site.

There is abundant evidence of simple ostensive modeling in the *Alvin* tapes made during early dives to the vents. Most of the researchers knew so little about the vents at this point that they were struggling just to describe to themselves and one another what they were seeing. Over and over again one can watch researchers on the dive tapes attempting to draw in their own minds simple conceptual maps of the area under study before collecting samples, setting down experiments, or taking measurements. This simple form of first-stage alienation was where they all had to begin their research.[19]

Usually the vent researchers began by identifying familiar touchstones in the vast ambiguous landscape around them. They slowly started to include more observations into their mental maps, extracting a crude but coherent image of the research site itself from the confusing rush of perceptions they were experiencing. The following example shows how difficult even this process could be. Scientists and the pilot of the sub are clearly not sure what they are seeing. They are looking for a hole full of bent worms that had been found on an earlier dive. The problem is that they keep thinking they have found it. The site is where the scientist and the pilot decided to put the pump (in the last quoted dive tape). The hole was named after the scientist who chose the site. Here I call it "Arlen's Hole."

#1: Yeah, okay, and that's where Arlen's Hole is.
#2: Where the weights are?
#1: Yeah, you—because I left them there.
#3: Okay.
#2: Isn't . . . wasn't the hole full of clams?
#1: Couldn't really take a look down to the bottom of it.
#2: Hm.
#1: No, I don't think it was . . . it was a . . . I don't think it was full of clams.
#2: H'm.

#3: I see a smoker over there. See the white smoker?

#1: Yes.

#2: Looks like [inaudible] tucked away.

#1: Let me see it . . . for later.

#2: There's one way over, and one just . . .

#3: Yeah. Do you see the one way over there?

#1: Yeah.

#2: It might be Arlen's Hole there, right where it is. Right to the right of the weight, right out here.

#1: Oh yeah, but here's a big hole too. You're going right over one.

#1: You can see the shimmering water. We're pretty close.

#2: Here's a couple of worms . . . instead they were bent in . . .

#1: Bent right over the hole, could that be it? If I described it to you . . . they were almost laying down on the [inaudible] bottom.

#2: All right. This is it right here. Coming up. [Inaudible].

#1: Yeah, that's the one.

#2: Is that the tube worms [inaudible]?

#1: Yeah, that's it.

These scientists clearly are trying to use an ostensive model of the vent site to find their way around the area. They want to do this because they are trying to locate a good place for setting down an experiment. The hole with worms in it has been recalled as a good location, but the researchers are having trouble finding it. The trouble is not so much a navigational one as a conceptual one. How can they distinguish this particular hole from any other? And does it really have any particular characteristics that make it more or less attractive as the site for setting down experimental instruments?

The scientists in the submarine made a model of Arlen's Hole and its environment, using a kind of question-and-answer conversational form that Knorr noticed among scientists who were studying pictures.[20] The similarities should not be surprising, since photographs, too, are kinds of ostensive models. Scientists in both situations are using ostensive reasoning, trying to find a shared vocabulary for the model, selected so that the model could guide useful perception and analysis of the object.

During one long dive tape that I studied, the scientists traveled over a large portion of the vent area numerous times in order to do a photo survey of the animals there. The photo survey itself I will discuss shortly, but for now what is important is how the scientists modeled the environment they were traversing in order to make the survey. They had to decide how to "fly" over the seafloor, using landmarks to identify a boundary around the survey area. They started to name rock "chimneys" that stood out in the landscape; they became familiar with a scarp that ran for a long distance along the side of an active vent field; they distinguished and saw patterns in the varying densities of organisms in the clam fields and used them to recognize the areas they wanted to survey; and with these landmarks they outlined a "field" for the study. When they wrote or talked about the area they surveyed, they described its location in precise navigational terms; but in the submarine they worked from an informal conceptual map they developed while looking at and talking about the seafloor. They made themselves "familiar" (as we say) with the area by reconceptualizing it as a map, making it an element of their culture with which they could think and in which they could move with confidence.[21]

Second-Stage Model Building

So far I have treated ostensive modeling exclusively as a conceptual activity, but it can also include the manipulation of physical objects. A stained piece of tissue, a slice of rock, or a culture developed under laboratory conditions are all ostensive models. They are selected to represent nature, but they are also manipulated to highlight some features rather than others. They no longer look the way they do in nature. They are made more useful for comparing to similar objects that also have been made part of the culture of science, through these alienating processes.

Researchers can create even simpler ostensive models of patterns in nature by collecting samples. Bits of rock, some organisms, or some water samples taken out of the deep become representatives of the sampling site, and, since they are not chosen randomly, they present selective features of that environment. In their combination of selectiveness and representativeness, they are more accurately called models of nature than samples. They are also models in another way. Once they are taken to a ship, lab, or piece of experimental equipment in the ocean, they become elements in the culture of science. Something is lost in the process. It is not just that fish will sometimes explode coming to the surface because the air in their swim bladder expands too much, or rocks may lose some of their features as they are broken into chunks that can be lifted onto a ship. Changes such as these are tolerated as long as the information lost in the process is not deemed essential to the model desired from the sample. What is lost is the overwhelming complexity of the site from which the specimen is drawn. The representative specimen reduces the site to a manageable object, alienated from nature to serve as its model in the culture of science.

When scientists sample from the deep sea terrain, they usually do it collectively rather than individually. This is part of what makes samples part of a culture rather than a personal construction of nature. The divers on the *Alvin* talk to one another about what they are seeing, what samples they want to bring back to the surface, and why these samples could be important. Their conversations in the submarine resemble the shoptalk of surface labs.[22]

#2: It seems as though the . . . the galatheids [crabs] are slowly approaching tube worms.

#3: Uh huh.

#2: But, uh . . . inside of, sort of . . .

#1: Which kind of crabs, now? [PAUSE] I only have galatheids out here but I see some [bythograea?] in the distance.

#2: There's a galatheid . . . that does exactly the same [PAUSE] slowly moving towards the tube worm. . . . There's these small tube worms here, a whole bunch of . . .

#1: Uh huh.

#2: They've. . . . Near here, they've sort of attacked. . . .

#1: Yeah, we used to have . . . I don't know that we've really documented an instance where a crab has really attacked a tube worm as it stands and killed, you know, and killed the whole thing.

#2: Here they are lying flat on the ground, so they can really march onto it.

#1: Yeah. Yeah.

#2: Next to your marker B, that is.

#1: Uh huh. [PAUSE]

#2: Look at the pink fish flat on the ground behind the dead mussel . . . and I'm [inaudible] looking . . . but it is not moving whatsoever, the tail flat on the ground too. And I'm so sure it is one.

#3: Huh, huh, see.

#2: Must be dead . . . but no crab on it.

#1: Have you taken a picture of it?

#2: Yeah. It will be hard to see on the picture. [PAUSE] Now there is a nice pink fish now . . . moving between the crabs.

#1: Huh.

#2: There is . . . the other . . . the other one must have gone through there. They rest . . . if they are lying on the . . . ground. I haven't seem them because they aren't moving.

During this and many other conversations among the divers, scientists "practice" doing science with one another, treating the submersible as a backstage area. They negotiate a reality from their observations at the vents, a model which they can take home. The social nature of the models is an important element. For in order to be placed conceptually within the world of science, they must move from individual minds or hands into a group. And this process is visible at the research site.[23]

Third-Stage Model Building

Models have had other uses in making discoveries at the vents. Comparing new observations to old descriptions has allowed scientists to locate new features at the vent sites. The scientists call these "discoveries," although many of the finds have been so common that it seems a bit presumptuous to call them discoveries. They probably should be called "normal" discoveries to distinguish them from "major" discoveries. The latter are research findings that contradict and/or fulfill theoretical prescriptions, giving scientists some good object with which to evaluate or formulate scientific theory (or for "thinking with" in the anthropological sense).[24] Normal discoveries are something else. They do not inform scientific theory or affect thought. They alter the descriptive details with which scientists theorize, but they only elaborate existing and accepted knowledge, not push the boundaries of thought. Here is an example from the vents.

> #2: [To the mother ship] Uh, does anybody up there have a description of the twin black smokers? This thing, uh . . . you only see one distinct black smoker that bears a narrow chimney about 6 inches diameter and about 4 or 5 feet tall, and then there's a clump that looks like some inactive chimneys next to it.
> RADIO VOICE: Roger that, Hal, Stand by. [PAUSE]
> RADIO VOICE: Alvin, Lulu . . . Hal, this appears to be a new smoker. Stay in that position right now while we get a fix on it.
> [The divers discuss how to use a video camera, while they wait for a message from the surface ship.]
> RADIO VOICE: Okay, Hal, go ahead. We were trying to get a good fix before we got back.
> #2: Yeah, I'm still waiting for a description of your twin . . . smokers. [PAUSE]
> RADIO VOICE: Okay, Hal, can you give us a description of the smoker before you leave and before we steer you back to the clam [field]?
> #2: Okay, black smoker. Probably 10 feet high, on about an 8-inch-diameter chimney . . . near the top. About 4 feet down from the top there's more black smoke coming out of the same chimney, just another exit point, maybe two chimneys right next to each other, but it's a very narrow chimney. The whole structure's very narrow. So you've got black smoke coming out of the top and out about 4 feet below that. About 5 feet to the right is a large . . . white smoker pump. I don't know how wide it's looking. It looks almost inactive . . . a lot of vent fish in it, a lot of crabs but not ventimentifera . . . and I don't see any hot water. Over.

A new black smoker goes onto maps in this area, but there is relatively little excitement about it because the phenomenon is familiar. The new discovery does not seem to get scientists closer to understanding the vent fields and their life courses. But it does make the value of any such explanation a little greater by expanding the physical area that scientific explanation can help to illuminate. Thus it modestly increases the power of science.

Fourth-Stage Model Building

Research may be facilitated by low-level ostensive modeling of nature, but it is often deemed incomplete or second class if it does not address theoretical issues in science. What constitutes theory and theoretical awareness may differ considerably from one subdiscipline to another in science (and in oceanography), but it always means some "big picture" of the natural world (usually drawn, as Galileo said, in the language of mathematics).[25] Researchers have dialogues with theories as they work. One man put it this way:

> I guess on the most conscious level [while I am doing research] I'm observing and recording as much of what I think is relevant information as I can, keeping an eye on the big picture and getting ideas [about] where it fits in, so that if I see something [important] I don't just go skimming right on past it. [I keep] the big picture in mind so that a decision can be made to go this way or that, decide where the hell [we need to go next]. And then [there is] the synthesis, first-level synthesis, that comes up when you're still out at sea, putting several dives together or several days' worth of data and making the map [from what you have learned] and figuring out what to do next. Then there's sort of a [secondary] synthesis that, based on observations, gives a view of the bigger picture; and then finally I'd say the fourth level [of analysis develops] in the next year or two later when you're back home putting together the data from the whole cruise, seeing how it impacts [exact scientific] ideas. [Interview with a senior scientist at a major university]

In theorizing, scientists use a kind of referential thinking, but one highly absorbed in the language of science. Theoretically sophisticated work refers less to the object of study than to a large complex of scientific theories and models that constitute the "big picture." What it means in practical terms to have the big picture in mind is harder for scientists to describe than what happens when researchers do not know what they are looking at or even what they should be looking for.[26]

> A classic example, classic example. [We dove with a submersible that] could only take one scientist, and they're diving and diving and diving and we were required to take [one] scientist or that wouldn't permit us to come back [to dive in that area]. Well, finally, all right, we let this guy dive and he was looking at my photo albums. Pillow lava, etc., getting prepped. Well, when he gets

in the submarine he goes down, he makes the dive, he comes back up. What'd you see? Oh, pillow lava. They process the samples. Many months later, many months later, a scientist [is] visiting a colleague in Brest, a very dear friend. And [my friend] has got all the samples that have been collected on this trip that I was on and no one else was on. They concluded that it, you know, it was a nice trip. Okay. 1978. And [the visitor] who's a specialist in sulfides is walking along the shelves. What's this? [My friend] says I don't know, it's a sample we got, I think it's altered basalt. He said, "Oh? Can I have a sample?" "Sure." Took it home. The discovery of polymetallic sulfides. That's how it happened. Because the [researcher] who came across the extinct chimney didn't notice it to be odd. And that great thing became the cover of *Nature* magazine, and the great discoverer of polymetallic sulfides didn't know what he was doing. And I can give you lots of other examples. [Interview with a senior scientist in a major research institution]

Knowing how to recognize evidence as significant to other scientists depends on knowing what it is that everyone is supposed to know. This is the social dimension of knowing "the big picture." Only then can a researcher notice an anomaly in scientific models or recognize where details of the familiar picture are not properly understood. Part of being a first-rate scientist consists of taking on the "big picture" as a personal responsibility, learning its details so well that the researcher begins to "feel" where researcher projects work and where they do not, where new models fit extant theory or where they challenge them.

You know, you look at something [like a geological formation on the seafloor] and you see a lot of them [using sonar pictures], and they have very systematic relationships and you just know they're important. You know, it's just instinct, I guess. Of course other people looked down and said, it's just a small transform fault. The same old thing we've been looking at for years, it's just smaller. And some people still think that. But it turns out that it is new and different, and once you try to explore how you can explain them and the physical principles involved it really starts to take you into understanding spreading centers a lot better. So that . . . these are where spreading centers end and another one begins and by looking at these . . . sort of, edges of spreading centers, if you will, you can sort of see into them . . . get a much better view for how they normally operate. We weren't the first people to map these things, but I think we were the first people to recognize that you can use it to really learn a lot new about the seafloor. And that's still an embroiled debate. Some people still think that they're nothing new at all, but it's sort of funny that the same people that argue that they're nothing new are sort of being drawn into the idea that, wow, we're really finding out a lot more about the seafloor than we ever did before by looking at these things. [Interview with a senior scientist in a major university]

Scientists like this one see themselves as redrawing the theoretical structure of their subfields by arguing with extant versions of "what everyone
knows." In this process, they *use* natural forms to serve their interests—
as alibis for exercising intellectual power over cultural images of the natural world. Nature becomes a resource to be employed in struggles for
domination within science and in the struggles by scientists to dominate
(at least conceptually) the natural world. They reach the fourth level of
alienation of nature to make it a source of social power. In the example
above, the scientist has drawn attention to a neglected feature of the
seafloor, is doing his best to make it seem important to explanations of
seafloor spreading, and is using his observations to generate a controversy to enhance his reputation in the field. His theories challenge conventional wisdom and at the same time indicate methods by which geologists can learn more about the seafloor. He uses mathematical
calculations to model the movements of the earth's crust in these areas
that would yield these features. In this way, he conveys a way of thinking
about the subject, not just his solution to the puzzle of why some patterns
in the rocks on the seafloor could exist, and he makes his science serve
his community by challenging his colleagues to think the way he does.
Even if they end up not believing him, they have to make their thinking
more sophisticated by responding to that challenge.[27]

What the researchers I studied said was most characteristic of the *best*
science (not just routine science but good science) was that the researchers "had perspective." A man discussing non-vent work described the
problem of lacking perspective in the following way:

> Now, when you speak of comprehension, you see [research projects] do not
> necessarily involve comprehension. The very penetrating person may also
> comprehend. I mean, there is wonderful comprehension. Like Maria Mayer
> suddenly realizing that one day when she was waltzing with her husband that
> the reason two electrons could occupy the same orbit was that one was rotating
> one way and one was rotating the other. They were spinning around each other.
> . . . But there's the other person that makes no attempt to comprehend. He
> merely discovers deeper and deeper and finally discovers an equation, ex-
> presses something, and now the expression gets his name on it; the equation
> gets his name on it . . . Now, going to the other part of the spectrum, the low-
> caste science, the guys who catalogue may have comprehension also. The fel-
> low who tries to understand what the universe can be used for, he may have
> comprehension. An example here is Carnot, who was an engineer and inter-
> ested in steam engines and who dug out of nature the so-called second law of
> thermodynamics—better than the man that was eventually given credit for it.
> . . . Let me tell you a little story [that illustrates lack of comprehension]. It
> was a few years ago. This story is not hypothetical; it's true. There is the disease

. . . which produces the symptom elephantiasis. It's caused by a mosquito biting people and injecting a little wormlike thing into them; it's a malarial worm and it grows in there and invades the blood in great numbers and gets finally so numerous that it stops up the lymphatic system and the lymphatic system keeps expanding to try to keep the circulation and you get these giant, distorted structures in arms and legs and so forth. . . . Well, a few years ago they developed a chemical treatment in which the disease could be eliminated so that the worm would be eliminated from the blood stream, the microscopic worms, and so they decided to test it on a small island in British Samoa. And they did, they treated the whole populace. Well, now it became clear. It was never a human disease. It was a disease in mosquitoes. It was very dangerous for mosquitoes to bite a person who had it. They were almost bound to die. And so [when the disease was eradicated], the mosquito population went up by a factor of 600.

. . . Then there are just plain stupidities, you know. Two years ago I went to a scientific [meeting] and I got in on this paper [by] this man [who] for twenty years has been conducting with hundreds of people and great cost, an experiment, a study, to try to discover why the more valuable range plants like clover, alfalfas, and trefoils, leguminous plants which are high in protein, have been dying off, versus the grasses and bushy things. . . . This is highly leached soil, lots of rainfall, I asked him, what are the trace elements in the soil? What trace metals exist in it? I asked him and he turned around to me and he said, "Why should I know?" and now I said, well, it does seem to me that one trace metal that might really be vital to your interest and that's molybdenum. . . . He did not know that metal is an absolute necessity [to legumes for breaking down nitrogen]. [Interview with a senior scientist in a major research institution]

Without "perspective" scientists can be lost in their own measurements. They cannot understand what they see or show its importance to others. In other words, they cannot generate meaning (or credibility or any cultural capital) from the data. The best scientists can add something to the culture of science because they have a broad knowledge of science so well internalized that they can feel discoveries intuitively and find ways to make scientific arguments from data.

Through this process, scientists expand the domain of science; they make discoveries that extend its reach to new areas of the natural world; they expand the theoretical and technological tool base available for doing research; and they make claims to a conceptual mastery of the natural world that is the basis for their social power.

CHAPTER NINE

Directing Scientific Discourse

WHILE IT IS CERTAINLY a source for the empowerment of science, alienation of nature in itself is not adequate to explain the power of science and its attractiveness to the state. More is needed to explain the fact that scientists have maintained control over the intellectual development of science in spite of their fiscal dependence. How do scientists make nature a resource for *science*, and not just for the state? How can they maintain control of their resources, if they are so valuable?

One part of the explanation must be that science is an esoteric knowledge system, which, no matter how valuable to outsiders, is difficult for them to penetrate and use.[1] It is not just that scientists write incomprehensible prose or propose theories whose applications are too obscure for others to see. If this were the whole story, the government could easily hire some "translators" from the pools of graduate students without good research positions. They could be like the people who write computer manuals, translating from a technical to a more common and comprehensible language. But they could never solve the problem. Scientists are involved less in generating information when they do their research than in generating controversies, struggles over data, and interpretation. (Science is dialogic, not monologic.) Interpreting these battles is an esoteric task indeed, for it requires technical, social, and political analyses. It is in fact one of the central jobs of professional researchers for positioning their labs' work strategically in a labyrinth of battles.

The state does have some resources for making sense of science. The rotors at NSF who have recently come from active laboratory research are certainly good translators of scientific debates, but they often have trouble making sense of struggles outside of their special areas of expertise. Moreover, they are not situated in the applied program where the problems of ongoing translation are most acute. So the conflict-ridden world of science remains something of a mystery to outsiders.

Bruno Latour helps to make sense of this world of research when he distinguishes between established science, which consists of principles on which scientists can agree, and science-in-the-making, which is fraught with controversies about the facticity, meaning, significance, and validity of any so-called finding. The science found in research labs is clearly of the latter type. It is part of the contentious world of intellectuals that Bourdieu[2] describes so unflinchingly. Moreover, this combative and contested science is what the state supports. It is not the idealized science of the past but the chaotic and petty science of the present in which findings of one laboratory are torn apart by their competitors. Since in this world the truth-value of even empirical results is often contested, trying to make scientific "truths" useful to the state is almost an absurd notion.[3]

Faced with this turbulence, agents of the state must attend less to translation of research results into practically useful forms than to the mechanics of the funding process itself. They must consider how to envision the outcomes of controversies so that they will spend funds on more "winning" ideas than losing ones; and in the scramble among different governmental agencies for money, they must figure out how to legitimate such a turbulent and frustratingly inconclusive activity as exploratory basic science. To help them with these problems, they call on the scientists who are engaged in important intellectual struggles to interpret research and articulate its value for the state.

Scientists, of course, are not engaged in intellectual combat in order to be obscure to the government or to appear like an unruly mass of eccentric and highly educated people; they are engaged in struggles for more sensibly programmatic reasons. Their fundamental charge is to make the natural realm more accessible to human understanding and manipulation, and this is discharged by a struggle with the natural world to increase the domain of science. Nature is not set out conveniently to allow scientists to find all the data they want easily or to be easily modeled with simple systems. So, doing science is itself a struggle. But there are also struggles within the social world of science among scientists who want to be the ones who are extending the domain of science most visibly and significantly; scientists arm their labs with techniques, problems, and personnel to engage in a kind of intellectual combat. In spite of the apparent tranquility on the outside, it can be quite bloody combat. There are territorial struggles among scientists dealing with more or less the same problem with different techniques and with different assumptions about the ways to solve the problem. Reputations are made in these territorial battles so they are played very seriously, even when they are more friendly than ferocious.

Ironically, then, struggles in science are a major source of power for scientists, their laboratories, and the institution of science itself. Success for scientists in the system (or for the signature of their lab) depends on whether their work becomes, first, "important" enough to be controversial, and later, powerful enough to prevail in these controversies by knocking out alternative approaches. It is hard for others to use the science emerging from these controversies without engaging in a battle to legitimate their moves, so insiders, who are trained in these fights, fare better than outsiders as translators of science. Thus scientists maintain control over their "information" and sustain some autonomy for their work.

The social world of science is an arena for intense intellectual rivalry in part because of the deep conceptual interdependence of scientists. In this age of specialization, scientists make sense of their findings by reference to theories and data from other research they cannot and do not know firsthand. They not only use a high degree of tacit knowledge that they acquire during their education, reading journals, attending meetings, and thinking about their research; they develop stakes in the work of other scientists when the latter bears upon the legitimacy of their own theories or findings. They join together in controversies to rip apart others or defend the work of sympathetic colleagues as a means of defending their own research strategies. The major currency of this game is reputation and the credibility of scientists' voices in their fields. But the seriousness of the game also derives from its relevance to the distribution of more tangible resources. What other researchers are saying and what they are finding can be very important to how, for example, funders will read the value of a research project or editors of a journal will judge the significance of a research paper. Because of their intellectual and political interdependence, scientists are intensely aware of what others are doing and how it reflects upon them.[4]

In this world of intellectual politics, having power (as an individual scientist or lab) means directing scientific discourse—and through it the practice of researchers. Defining legitimate research and theoretically significant issues to address with it are means with which scientists exercise the power left to their community by the state. Using it effectively is the ultimate prize in science, and there are a number of ways to do it. A riveting finding can do the job, like the discovery of the structure of DNA. The discourse in biology was permanently changed by it. Being the successful proponent of a new theory can also do it; take the example of the "color" theory of quarks that Pickering describes. One could also include here the theory of plate tectonics in geology as a recent breakthrough that has steered scientific debate and activity in new directions.[5]

Besides these dramatic conceptual shifts, there are subtle and less purely intellectual ways in which scientists can change debate. They can systematically discredit a line of research, without drawing particular attention to their own; they can develop a new research technique that allows a new line of research to be pursued; they can be part of a group or network of people who dominate a field socially (and through their association also intellectually); they can take science to the mass media and shape its public image in a way that gives them power within their own subdiscipline; they can run journals and scholarly societies and through them affect the reward structure of science; or they can develop policies for funding agencies that send research in new directions. The list could go on and on.[6]

This quality in scientific practice seems to contradict directly the image of scientists as dispassionate observers, voiced by Merton in sociology and, more recently, Keith Thomas in history. Thomas, for example, has written that around the time of the so-called scientific revolution, Europeans began to relate to nature in a more depersonalized fashion. Before the period, people thought of the earth as a domain given to them by God. In their relationships with plants and animals, people were driven by a vision of nature as a kind of servant, designed to serve the needs and whims of human beings, whom God had made sovereign over them. What replaced this point of view was a more detached and disinterested attitude toward the natural world. Scientists asked more questions about how nature *worked* than what could be *done* with it (particularly after Baconianism died out).[7]

What is perhaps ironic in the present (and perhaps also the past, although I have no data on it) is that scientists are among the groups with the *least* detached and depersonalized relations to the natural world. They have territories; they are wary of intruders (including both non-scientists and scientists); and they are driven to delight and despair by the results of their experiments, sometimes finding nature an ally to use in their interests or an enemy ready to frustrate all their attempts at doing their work.[8] Passion and sovereignty more than detachment characterize the way scientists relate to the natural world (or their traces of it) in the daily life of laboratories.

> When I came to [Big Lab] there was a man who's still there. . . . He worked in [the Pacific Ocean]. He welcomed me very kindly. But one thing he wanted to get clear. The Pacific was his [territory]. Since I had been hired to work in the Gulf of Mexico, I didn't think too much of it. But it seemed an awfully big chunk he had bitten off for himself. But then I could have the Atlantic. So, for many years I worked through an East Coast institution. [Interview with a senior scientist at a major university]

Clearly, the reward system in science favors those who successfully dominate the minds and research techniques of others, and this situation prevents dispassion from being as decisive as the norms and myths would have it.[9] Researchers engage in a kind of micropolitics to protect and promote their voices in scientific discourse. Through this process, they not only add passion to the process of research; they affect the images of nature that scientists draw upon when they try to conceptualize their individual projects and problems. The expansion of the domain of science is accomplished by the actions of thousands of researchers engaged in intellectual strategies and the groups they organize (formally or informally) to serve conceptual/empirical movements.

> It was a long development in this hot-spring program, starting fifteen years ago. Then there were various chemical and geophysical anomalies in the ocean and we found a problem that couldn't be explained by any obvious process. And then the consensus began to develop among theorists at [Big Lab]. Almost everybody involved in the original hot-spring work was [from Big Lab]. It was a fantastic crop of graduate students who went through [Big Lab] in the '60s and early '70s in all fields. And, you know, we knew, of course, plate tectonics. . . . And the consensus developed in talking around about all the things worth doing out there, and the consensus was [that] there had to be hydrothermal activity on the seafloor, the ridge axis. But the question is, was, how do you find it? . . . And then a big consortium of us dove over that thing and sort of sat on it, and discovered the Galapagos [hydrothermal vents]. [Interview with a senior scientist at a major university]

Extending the domain of science is not just a matter of getting new information to add to a stockpile of knowledge; scientists, cohorts, disciplines, and other groups carve out territories of research, and then devote themselves to increasing the significance of the territories to other scientists and maintaining some power over their territories. Research groups engage in competition with other laboratories, using their "signatures" to make territorial claims, and criticizing the "signatures" of their adversaries to enhance their positions in the zero-sum game of finding "truths" about their common territories. Disciplines and subdisciplines also engage in territorial rivalries to gain or sustain authority over territories and increase the value of their established territories.

To understand patterns of research over time, then, one cannot look to funding policies of the state alone, in spite of the massive dependence of science on the state. One must look at the multiple voices in science vying for attention and for territories to claim, trying to take advantage of the interdependence of scientists to dominate the thoughts and research practices of others through their own work. They are the ones that attempt to become authoritative and supply the voice of science with new forms of discourse and new sources of legitimacy.

DISCIPLINARY RIVALRIES

One might think that the territories of disciplines (if not subdisciplines) were stable enough to engender rivalries or conflicts, but two factors work against it. First, the relative importance of a given discipline in science can change over time as theoretical and empirical breakthroughs draw attention to one field or the next. The discovery of DNA and the resulting research and bioengineering have given biochemistry new prestige and intellectual life. Similarly, during the 1960s and 1970s the research on plate tectonics in geology gave that discipline a boost in prestige. Dramatic changes in relative power occur infrequently, but small swings in the balance of power are common enough to be almost routine. They stimulate or reinvigorate interdisciplinary rivalries that may be only ritually observed during stable periods, when the exchange between disciplines and the boundaries between them are not problematic.

Secondly, the stability of these boundaries is affected by the kinds of interdisciplinary work done in a given period. The more that scientists from different disciplines depend on one another conceptually, empirically, or politically, the more rivalries can develop. It is elementary sociology to point out that people use social conflict as a way of marking boundaries, and where there is a great deal of interaction across the boundaries (so they get fuzzy) social conflicts may arise to help clarify the lines of demarcation.[10] At the same time, interdependence of researchers from different disciplines also breeds animosities because the boundaries do have significance, even when they are being eroded. Different kinds of expertise are cultivated in disciplines. Biologists interested in understanding how deep ocean creatures interact with their environment may need to know characteristics of the chemistry of seawater, which they cannot find out for themselves. They are clearly dependent on colleagues from chemistry to perform the experiments they need because they do not have the expertise to do it. This dependence is based on *incapacity*, so it can and often does aggravate interdisciplinary antagonisms.

Researchers from one discipline who routinely read in another literature or often collaborate with colleagues in other disciplines may also develop quite disdainful attitudes toward the others because the rules for doing research and gaining esteem vary among disciplines. They experience an irritating "culture shock" when they find their research practices criticized in ways they find not wholly legitimate and/or difficult to anticipate, when they cannot make what they take to be a critical research question seem important to colleagues, or when they end up assisting with experiments they do not understand. This last pattern can be seen in the following exchange between some geologists and a pilot on an *Alvin* dive who were asked to collect clam specimens and install a crab trap for some biologists.

#1: Okay. Well, for the moment . . . just land, you'll see in a minute an incredible number of clams coming up. A big field of them. Okay? We're gonna land here and just take all the clams and get that over with. You can pick them in one arm, I'll pick them up in the other, and we'll just do that and then we'll move over to the sulfides.

#3: Some of this stuff is covered with black . . . black zinc [inaudible].

#1: I beg your pardon, John? John. Andy. Andy, you are Andy, aren't you? Just land in that field where our bulk manipulators can have a heyday. Okay?

#2: Yup.

#3: All the pillars are covered with black, I mean zinc sulfide. That could be manganese, could be zinc . . . different color. . . . [PAUSE] We need to find live [clams].

#3: You wanna take the live ones?

#1: I'm not [sure]; I'd like to take both but mostly dead. [PAUSE]

[Completely garbled conversation]

#1: Okay. Hey, you know, if it's just as easy to pick live [clams], pick them up because we can uh . . .

#2: We can kill them.

#1: . . . kill them. Make them dead ones, ha-ha-ha.

#3: The live ones are [inaudible].

#1: And I don't see that other crab, so let's not put that crab trap down yet. [PAUSE]

#3: We pick . . . we should pick live ones because then the shell is harder, no?

#1: Yeah.

#3: And we can get into the meat a little better.

#1: Yeah, I'm gonna pick up just live ones. And there are so many of them.

#1: Give me the . . . give me the crap, the claw back. We don't want to obliterate every subject matter, ha-ha-ha.

#2: Can we put the trap down now?

#1: No, we won't put the trap down until we see the right crabs. And we'll do that while we're working on sulfides. Did you give it back to me?

#2: Hang on.

#1: Okay.

#2: Okay.

#1: Okay, thank you, sir. Got her. [PAUSE]

#2: Think I'll read the *Playboy* or something. This is boring the shit out of me. Down in the dumps.

#1: There you go . . .

[Completely garbled conversation] [PAUSE]

#1: Ha-ha-ha-ha-ha. It's like grabbing a bar of soap; you just squirt it out. [PAUSE]

#2: Okay, it started bleeding.

#1: You crushed it, you turkey.

#3: Bleeding?

#1: Don't crush them.

#3: The sharks are gonna come. Maybe . . . only bigger predators would come for the blood.

#2: Crabs.

#3: The big fish are waiting outside [inaudible].

The frustration of these researchers in collecting the clams results in part from the practical difficulties in collecting specimens with the claw of the submarine, but there is more here, too. They don't really know whether to collect live or dead clams; they find the work alternately boring and frustrating because it lacks meaning to them. Apparently, it is not interesting to them and presumably cannot be used to promote their own thinking or careers; they find the patterns of life in the deep mysterious and perhaps even menacing (when they talk about sharks); they want to get the biological part of the dive over with so they can proceed to do the kind of work they feel competent to do; and even while they are doing the biology; they tend to revert back to discussions of the geological features around them. Later in the dive, when they are finally able to concentrate on their geological research, they will still have problems collecting specimens, but they will know what they are doing and that will make a dramatic difference.

On these tapes, the researchers were treating biological work as a kind of "dirty work" (in Everett Hughes' sense) that was getting in the way of their other (less dirty) research.[11] They assume a hierarchy between the disciplines, but they do not defend it very vigorously because it is not being threatened by their cooperation with the biologists.

This kind of mild antagonism toward doing some biological work for colleagues during a large, interdisciplinary research cruise is nothing compared to the ridicule that another geologist inflicted on biologists in an interview, when he said that finding the vents provided an opportunity for a moribund biology department to get some funding.

> What happened was, we thought a biologist should come on the original Galapagos expedition, and they said, why? We said, you never know. Then [when we went to the vents without a biologist] we found the organisms. We sent a cable back to [Blue Lab] saying we'd found clams and amazing-looking worms, and blah, blah. We were geochemists and geophysicists, and we didn't know what we were looking at. We didn't know what they were. Big clams and big worms. And I got a formal cable immediately back from the director of [Blue Lab], bureaucrat, saying that these organisms have to be preserved. They are the property of [Blue Lab], since there was nobody funded on the expedition to do biology, which was legally right, you see. Basically, what he was saying was, hey, wow, there's a moribund biology department that can't get any money. Funding in marine biology is nothing, you know, nothing. All of a sudden it looked like there was going to be a big, big chunk of money for those

guys. And of course there was. Hundreds of thousands. I mean, big to the biology [people]. . . . Biology put far more money into stuff than the rest of oceanography, probably by a factor of five. We [the geochemists and geophysicists] find [the vents]. They don't have to spend the money finding them. They just go and follow on behind us . . . I mean, finding them costs a quarter of a million dollars a month—the exploration program. Those guys just go out to the X we leave on the map and dive on it, man. [Interview with a senior scientist at a major university]

This man clearly resents his relationship to the biologists who studied the vents. He does not think about some abstract, ideal relationship between the fields in science that makes it good for scientists from different disciplines to build on each other's research. He sees the biologists as vultures seeking a piece of his kill. He acts territorially, and uses his substantial rhetorical skills to defend his territory. Implicitly, he was decrying the fact that the biologists were getting more attention (in the press *and* in the scientific community) for their hot-spring work than the geochemists and geophysicists who had predicted and discovered the vents. The resulting growth in the prestige of biology was making the hierarchy of disciplines in oceanography problematic in a way that threatened the researcher and evoked from him a strongly territorial response.

In another interview, a biologist illustrated what a difference this work was making for members of his discipline. He expressed his growing pride in vent biology and his sense of superiority to the geologists and geophysicists working at these sites. He criticized the value of the geological work and the research techniques used at the vents. He was a little hesitant to go too far in his criticisms, given the fact that marine geologists had been so deeply involved in plate tectonics theory that their power in science had grown enormously in recent years, but his willingness to be critical *at all* illustrates the extent to which marine biologists have been or have felt themselves empowered by work at the vents.

You know, much [of] geology is working out of a mega-theory; it's different from what we're talking about, certainly [in marine biology]. I think most of the vent work [by biologists] is just trying to find out how this system works on different levels and it doesn't constitute a mega-theory really. It's more a description of how this biological system, which is only very recently discovered, works. And it, you know, it has a lot of interesting things in terms of different properties or proteins and sulfides binding, sulfide which is highly toxic . . . But it isn't anything like, you know, the big [theories like] plate tectonics. I mean, partially I think, you know, this is what geologists might not agree with, partially the reason that they're interested at that level is that it's so hard for them to actually get in and analyze the way you can with biology. I mean, all they can do basically [at the vents] is look at the surface of the rocks and they can do some things that tell them something about what's inside but it's very difficult for them really to know. . . . You know, the time scale is so great they can't do manipulations. So, in terms of just a lot of sort of approaches that biologists take, their hands are tied. They can't analyze things in many cases in any kind of detail. [Interview with a senior scientist at a major university]

These kinds of criticisms across disciplines are widespread, fortifying boundaries during cooperative research by scientists of different disciplines and shifting rivalries as changes in the disciplines begin to affect the relative power of scientific fields.

INTRADISCIPLINARY RIVALRIES

While this kind of cross-disciplinary criticism is common, it is not nearly as frequent as the squabbling within a discipline or even a subdiscipline. You can find examples of this both during cruises and in the journals. At sea, oceanographers carve out physical territories, using equipment as markers. Collectively, scientists claim research sites for science with their transponders, pumps, samplers, and other instruments strewn about the seafloor. Individually, laboratories claim separate intellectual and physical territories. Other scientists are allowed in territories claimed by others, if they are helping with the experiment, but otherwise, they are supposed to leave the stuff alone. These implicit rules were made manifest when a scientist from one laboratory began to collect samples of tube worms on a site where members of another laboratory were taking photographs to assess the densities of animals. When this happened, the researchers on the submersible joked and laughed nervously about what would happen if the senior researcher in the other lab found out about it. He clearly would be furious since his counts would no longer reflect the natural densities of animals, and he would have to find a new site to survey. In this case, the territory was not marked with any standing equipment (which perhaps made the territorial claim ambiguous enough so that the researchers felt they could get away with their breach of etiquette), but their nervous laughter indicates that they implicitly recognized the existence of territories and their breach of the rules of territoriality.

Onshore, scientists try to share a limited territory and expand their territorial claims by dominating the work of other scientists. Lines of cooperation and conflict are drawn by groups attempting to explain more or less the same phenomena in different ways.[12]

This is a major discovery, and I think the whole major discovery is that animals can live in the absence of light, and on earth, there is none . . . well, certainly in a cave the plant life gets into the cave somehow. But in the deep [ocean], no plant life gets down . . . That's where [my competitor] made his big mistake. There was a group there at [Big Lab] that got together and discussed the idea, this chemosynthesis idea. [And they thought] there must be some other explanation of this. And they are all very good people, very clever. They wrote that up, and they sort of needed someone who had been on a cruise, so they got [my competitor]. The idea being that there is something else, a heat circulation, a hot vent goes up and the water circulates, and by doing this, it scrapes the bottom off of the particulants and carries it to the vents and feeds the mussels. That was the original idea. But the idea has flaws, mainly that if you get this, you exhaust yourself because there is no input, there's only a [circulation]. So there should be an end to [the food supply], but there's clearly no end to the mussels, to the vent fields. And some plumes [of hot water at the vents] are very small. Where there is most life, the plumes aren't that big . . . over the bottom. And then the water is cold and there is no circulation [or it's] so very small. [The food supplied by circulation] should be very soon exhausted, so you have to think of input from the surface, and calculate that in square miles of surface funneled down to a little clam field. They haven't done that. [Interview with a senior scientist in a major research institution]

This man is describing a rivalry between "his people" and another group at another research institution whose work opposes his own. In describing the work's analytical merits and deficiencies, he is doing much more than detached analysis. He is fighting to gain the greater voice in defining the ecosystem at the vents, that is, engaging in the politics of intellectual discovery.

Claiming that this researcher is using political skills to advance his theory is not to say that his ideas do not have merit. On the contrary, scientists fight for the ideas that they think have merit to try to insure that they will get the attention they deserve. In the daily practice of science, intellectual and political issues are deeply intertwined. We can see this more clearly by looking a little more closely at this particular controversy.

The political process of determining the relative power of two interpretations of the sources for food at the vent sites began when the scientist I quoted above was confronted with a prepublication version of a theory that was a rival to his own. He responded first by asking his competitor to reconsider his views on the new theory. But the other research group submitted their paper to a journal, where a journal editor, one of the power brokers within the field of science, tried to mediate the dispute. Here is my informant's version of these events:

And they wrote it up, and got [my competitor] into it, and he sent the paper to me. And I told [him], "Don't do that. This paper has no facts and no data.

It's all speculation. And it goes in the wrong direction." Well, he was sort of impressed by this other group, and they sent it to *Science*. I don't know who they knew. Well, [the people at the journal] sent it to me. And I had my review already in my head and I made all these points about why it is not a very good idea, and sent it back to *Nature*. Six weeks [later] it appeared, unchanged. No caveat or anything put in. I was interested in the other reviews. I was saying, "There must be other reviewers that think this may be so, and is it possible to hear the arguments? If the reviewers [must remain] anonymous, would you [at least] let me know what they said?" The editor said, "No, no, no. The other reviews are even stronger negative than yours. But listen. This is such a nice interesting area we thought it should be put to the public, and if you look at the issue of *Nature*, this is not a reviewed paper. It's under 'Views,' 'News and Views.' " That's not a real tribute. So I thought about it, and I said, "Well, you asked me to review the paper and you asked two others and then, if you, for journalistic reasons or editorial reasons, you don't want to listen to the reviewers, you publish it anyhow? This is sort of strange." And only then he realized and he apologized. . . . And now I'm not too unhappy about it because I can use it; this argument of course exists now in the literature, so I can refer to this. [Interview with a senior scientist in a major research institution]

By extracting an apology from the editor and demanding a voice in editorial policy of the journal, this researcher tries to shape scientific debate, using institutional as well as intellectual resources. His aggressive action is a move to claim and defend a territory. By printing the paper as "news" rather than as a "real" paper, the editor denied the seriousness of the claim made by the group who submitted the paper, but he did not discredit them altogether. That is why the scientist quoted above made his objection. The editor's decision to print the paper wrested control of the issue away from him. But in the end, when the editor apologized for printing the paper, the legitimacy of the scientist's claim to that territory was at least informally acknowledged.

This case is interesting because it brings to light details of how institutions like journals function as arenas for territorial struggles and as a place where the legitimacy of claims is (at least partially) decided. It illustrates the *collective* nature of these claims and their adjudication. The speaker suggests early that he tried to get his rival to *join with him* rather than with the group that wrote the article.[13] And to make his claim potent, he calls on the opinions of other reviewers of the paper and on the editor's actions.

Another interesting point raised in the quote is that authorship of an article is treated here less as a measure of who developed an idea than a measure of those who want to associate themselves with it. For scientific articles based on data gathered during large research projects, there are so many scientists involved that a wide range of people could legitimately have their names on the list of authors for a paper. Who winds up there is partially a political, rather than intellectual, matter. A group writing an article can decide to bring someone else in to work on it. Potential authors are faced with decisions about when and how to join these groups.[14] Part of forging an individual career and shaping the direction of a laboratory, then, is deciding when to join or break away from a group.[15]

While these decisions affect a scientist's stature in the field, what affects it most is the research initiated by his or her lab. This is where they try to innovate and gain the prestige to make others their followers.

I decided we should look at developing a technique which would let us look at oxygen, CO_2 and sulfide. . . . The way that people usually look at a lot of these things is to use labeled carbon. I think that's about what basically everyone else in the program uses plus some use labeled sulfur for some things. There are real problems with using labeled carbon. . . . What happens with a green plant is it takes CO_2 and makes it into organic matter. Okay, and all the classic work here . . . a lot was done with labeled carbons and 14. The only problem is that at the same time the plant is also metabolizing organic matter back to CO_2. So when you put in label, part of it's being incorporated and part of it's being metabolized and turned around. It can be fairly complicated to know what's going on, so it's just not an approach I've ever really liked very much, although it's been very powerful for some things. [So I] spent a year or so adapting a gas chromatograph to [read oxygen, CO_2, and sulfide].

. . . The blood of these animals [at the vents] often had very high sulfide concentrations, [because of the sulfides in the vent water], much higher than in the environment. And this led us to the idea that these animals have sulfide-binding proteins that were analogous to the function of hemoglobin with oxygen. Hemoglobin . . . one way to look at it is, it concentrates oxygen from the environment. And in the same way these animals, the tubeworm and the clam, have sulfide-binding proteins which concentrate sulfide in the environment which then is carried to the bacteria. So, we think the function is very analogous. . . . So that [idea] was dependent upon making a decision fairly early to put a lot of effort into developing a little different technology than anyone else had. [Interview with a senior scientist at a major university]

To make territorial claims (here about the way animals use the sulfides), scientists need to have their own approach to the problem. This approach is generally designed as a response to critiques of the work by competitors. In this case, the scientist has focused on measurement techniques (the problem with labeled carbon studies). Others sometimes pay more attention to interpretation of data than methods of collection or theories to reinterpret existing data. The point is that scientists continually find some way or ways to differentiate themselves from others to make territorial claims and to try to recruit allies to their work. This helps them to dominate an area of scientific debate.

STRUGGLE FOR PRIORITY IN THE DISCOVERY OF THE VENTS

Scientific arguments over proper procedures for doing research, appropriate interpretations of data, and the theoretical credibility of arguments might seem to flourish where scientific research yields ambiguous or controversial results. Thus we might expect that a straightforward finding, such as the identification of a new animal or the location of the first hydrothermal vents, might be simple enough to avoid these problems. But certainly the discovery of the vents was not simple. A number of people working together on an expedition to find the vents either claimed the discovery or had it attributed to them. It does not seem that one scientist stole the idea from another, although the idea of searching for the vents certainly seems to have been passed along a line of people, complicating the situation a bit. The primary problem seems to be that "ownership" of the discovery was made fuzzy because the discovery was a *group* effort.[16] This case, then, has a lot to teach us about the power struggles within science and the ways that they are played out in large research groups.

To summarize the situation briefly, those who claimed (or were given) priority in discovering the vents had done the following: (1) made the first record of the vents; (2) first proposed the cruise to find vents; (3) organized the cruise that was successful in locating a vent; (4) first saw the vents; (5) first recognized the sweeping significance of the finding; (6) made the first extensive visual survey of the vents; (7) first published about the vents; and (8) received the first massive media publicity about the discovery.

One of the things that made attribution of the discovery a little difficult was that although hot-water vents had been predicted by geologists before their discovery, no one had predicted the abundance of life near them (see fig. 6). Geologists had surmised that water was circulating through the rocks on some areas of the seafloor and was escaping into the ocean as hot water. And they had found some temperature anomalies in seawater near the bottom of the ocean, which they thought might have come from hot springs. That is why researchers were on the bottom of the ocean looking for vents. But the vents turned out to be much more interesting than anticipated. No one knew that there would be new and massive numbers of animals around the vents, or that interesting kinds and concentrations of chemicals would be coming out of the hot springs.[17]

It is also difficult to say who found the vents since what was found was so unlike what was expected. Did discovering the hot springs entail locating some place on the seafloor where hot water was coming out? Not

Figure 6. Hydrothermal vent. This photograph illustrates the dense life forms found crowded around the vents. (Courtesy of Woods Hole Oceanographic Institution)

exactly. No one seriously claimed that the first scientist to find significant temperature anomalies in the area where the vents were discovered was the "real" discoverer of these phenomena, but he certainly was the one with the first date indicating hot springs. Another geologist was closer to the "real" discoverer. He made photographs of the seafloor that included pictures of vents, so he was not reminded what he was looking at. The camera sled was not recording temperature, so he was not reminded of the hot-spring hypothesis in geology, and he was not trained in biology so the pictures of great clam fields did not mean anything to him. He had no idea how odd they would appear to a benthic biologist. He obviously *could have been* the clear discoverer of the vents, but he was not quick enough to recognize what he had found. That means that the only candidates for discovery came from a large group of scientists involved in a set of cruises designed to locate and examine hot springs because they had been predicted from geological theory. These people were mainly geologists looking for hot water, not biologists looking for massive quantities of clams and crabs, so in some sense they did not find what they were looking for, and this complicated the issue.

Just as conceptual issues complicated the discovery, so did the organization of the cruise. It is difficult knowing who really thought up and organized the cruise on which the vents were discovered. Two scientists had a large role, and this helped to set up a rivalry between the two. An expedition to search for and study the vents was first proposed to NSF in 1974 by a senior professor at a relatively small but growing oceanographic institution. This preliminary proposal was given favorable review at NSF, and the researcher was asked to send in a full proposal. But by 1975, when a full-blown proposal was written, the cover sheet indicated that an assistant professor at that institution was now the Project Director, and his senior colleague was at this point only one of four Co-Principal Investigators. By the time that preliminary studies had been done to determine where there was enough hydrothermal activity to warrant some *Alvin* dives and a continuation grant was sought to fund the dives, the original Principal Investigator had moved to another institution, and he was now only an adjunct to the main group running the study.

Although the titular head of the project was the junior scientist now named the Project Director, the senior scientist who had originally written to NSF still felt he was the "real" head of the expedition. He claimed that the direction of the project had originally been given over to his junior colleague because he knew he was moving, and could not take the grant with him. (It had been proposed to be administered through his former home base, and NSF takes the location of the grant into consideration when making funding decisions.)

The controversy over who was the real head of the project became significant only because the junior colleague, in part because he was the Project Director when the vents were discovered is often credited with the discovery. The senior scientist feels this attribution unfair. Sociologically, the interesting point is that by giving up control of the money, he also seems to have given up much of his claim to the territories staked out with that money. In losing control of the means of production, he also lost control of the intellectual "fruits" of the expedition.

The rivalry between these two scientists was exacerbated by the fact that they were the two researchers on the submarine when the vents were first seen and identified by eye. Who discovered the vents in the sense of seeing and recognizing them? This seems a simple enough question, but the answer is once again complex. The senior member of the rivalry put it this way:

And we were just incredibly lucky. We arrived in the area and the light was green. For one reason or another, the sub functioned the first time that I took the first dive—together with [the rival]. And [he] had a wonderful machine that was full of little blinking red lights that took three-quarters of the sub space developed by the GEOSECS group to measure the variety of properties of seawater. And out in front of the Alvin hung this tube and pumped seawater into [my colleague's] machine and it measured the pH and the salinity and the oxygen content and a lot of other things, and [he] was sitting involved with his back to his porthole, being completely mesmerized and little numbers would appear in the window and he'd dictate them in this tape recorder, which struck me as kind of silly since they also went on the tape in the machine and he obviously was always behind it. And we drove around, looking for the little needle in the haystack. How do you see warmer water, not a lot, just a little bit warmer—not two degrees but two point zero, zero seven five degrees—in the middle of all this water in the darkness and you can't see thirty-five feet out of the porthole. So we first went to the fissures, which we knew were there from the previous photographic survey, and cruised along them. Occasionally you'd get slightly higher temperature but you couldn't see anything. And then around eleven o'clock in the morning, I'll never forget it. This basalt is very black, and it really looks unkind, rather forbidding, austere. There were these white patches, like snow. They looked different from ordinary ocean sediment, which is also white but that's a different sort of salt. This was bright, brighter than white, and so the pilot saw it and I saw it and we turned back, and often when you turn back you can't find it again because your circle of vision is only from me to about the wall there. . . . But by sheer luck we cruised back around over it and we went closer and it turned out to be shell, clam shells, this size [hand gesture]. And we thought neither the pilot, who must have had over three hundred dives in his life, nor I, with far less but still twenty or thirty, had ever seen a clam shell that you could see with the naked eye on the deep seafloor. The largest one seemed so far from [inaudible] and so at the same point, and I got excited that I said [to the other scientist], "Look up! Look out your porthole; something odd's going on." Then we pushed a little further and there were heaps of mussels and the water took on a funny shimmer and it became cloudy and there were all sorts of clams, just masses of them, just all over the place. And you know the ocean floor generally is very barren. You see, every hundred yards, one animal that you can see that's really plenty, but here we got thousands [in] a hundred square yards. And [the other scientist] said, "Don't bother me. We're just beginning to read a higher temperature and the oxygen is going down." And so he was mumble, mumble, mumble, dictating the numbers and he wouldn't look out the porthole. We cruised over this thing several times, called up to surface and said get a good fix on us, we're onto something important; we don't know what it is. At that moment their computer broke down so they didn't get a good fix on us. Finally we went back to the surface because the eight hours were up and we came up and they said, "Well, did you see the hot spring?" and [my colleague] said yes. It was the temperature increase. He said it's very exciting. There is no oxygen in the water. And they said, "What did you see?" And I described [the fields of clams and mussels, and the shimmering, milky water]. And that seemed all unbelievable to them. So they turned to [my colleague] and said, "Did you see all that?" And he said, "No, I didn't see anything." And my credentials went down just like that, you know. He kept insisting. . . . And the pilot, who is a little devil, wouldn't say anything. He just left me telling this unbelievable story. [Interview with a senior scientist at a major university]

This researcher clearly seems to exaggerate for effect; the films made on the submarine during that dive indicate that they picked up samples of the clams to take back with them to the surface ship. Clearly, the researcher had these samples to support his case. But he did face a severe problem of credibility; what he saw was so new that it would not seem plausible to other scientists. How could anyone listening to his story believe the densities of the clams that he had seen in some areas? And since there was a broad area in which there were widely scattered clams, it is quite possible that his junior colleague had not seen some of the denser areas and merely reported seeing these more scattered specimens. Photographs from the dive described above were developed onboard and finally vindicated the scientist who seemed crazy when he described these areas as packed with living creatures. And other divers soon added new testimony to his voice. But in spite of this, he gets relatively little credit for the discovery in comparison to his junior colleague.

The problems of attribution in work conducted by large groups of scientists, particularly groups with scientists of different social rank (both institutionally and in terms of reputation), exist in part because there is so much work done by junior researchers that is funded in the name of more prominent senior colleagues and published with the senior scientist as first author. Reputations are so important in getting the attention of funders and journal editors that this practice has practical advantages. But it also makes it more difficult for the parties involved to make claims about discoveries and for others to assess the legitimacy of their claims. It is difficult to know when a senior scientist is properly considered the "real" first author of research proposals and analyses.[18]

Although in many cases this leads to increased advantages in publication as well as in grant getting for senior scientists, in this particular case publication etiquettes may have worked in favor of the Project Director and against the interests of the senior researcher. Many of the early descriptive papers that introduced the vents to the scientific community were co-authored by a large group of scientists, with the final Principal Investigator as the first author, so that his claim to priority (as the first to publish on the subject of the discoveries) was enhanced. In addition, he published in scientific journals and in *National Geographic*, so his name was associated with the discovery both in the public mind and in the scientific community.

Oddly enough, the man whose name is most frequently associated with vent research in the public mind is a man who had no crucial role in the cruise. He did make extensive photographic surveys of the vent sites, but what was most important was that he was the one who was able to attract the most media publicity to the cruise. He had already done work with *National Geographic* and had a comfortable working relationship with the people there. His name was already known to readers of the magazine, which made him the natural co-author for the pieces on oceanography published there. And he became the informal "star" of a documentary about the deep ocean, including the vents, which was shown on the Public Broadcasting System. He was important to the research to the extent that he was the most experienced with the submarine and was able to use cameras effectively to record what others were seeing; thus he contributed to the documentation of the dives, but his publicity was what made lay people associate the vents with his name.

All this might lead the reader to conclude that the priority issue was in the end quite simple. Scientists attributed the discovery to the Project Director, while the public attributed it to the discovery's publicist. But it is not quite that simple. Walter Cronkite took his dive to the vents with a different scientist; other researchers were interviewed on the "Today" show; and many of them were part of a range of local news stories and newspapers reports of the first cruise and a subsequent one. All this made the question of publicity a little more complex than it might seem at first glance. And even the status of the Project Director is problematic for the scientific community because he dropped out of the research world not too long after the cruise was over. His colleagues suggest that he was unable to handle his success and dropped out of normal science. All this had muddied the waters more. Researchers in this area seem reluctant to attribute priority to a man who is not around to receive the rewards for it, and the public is now less clearly focused on any one individual as the discoverer of the vents.

Some of the problem of attribution is simply an artifact of big science. When groups work together, how do you distinguish the contributions of each? Certainly, people do. But the grounds on which they make these distinctions tell us perhaps less about the individual scientists or their labs than about the competition in science that still underlies the cooperation necessary to make big projects work. It may also be that attribution is far less important to science than sociologists have contended. It may be that controversy here functions like controversy in other areas of science, drawing attention to those involved and thereby rewarding them all. Only where Nobel prizes or vast differences in recognition for researchers are involved do these distinctions really matter.[19]

SCIENTIFIC EXPERTISE AND SCIENTIFIC KNOWLEDGE

The complex meanings of any finding or definition of a problem depends on and speaks to a world of tacit knowledge that scientists bring to their research. However much scientists fight with one another, live in glorious isolation in their specialities, or define themselves and others through intellectual criticism, they also depend on one another to develop coherent images of the natural world with which they gain broader perspectives on their own work. It is this interdependence that makes it possible for researchers in one field to influence work in other specialities; therefore, cooperation is at the heart of the competition in science. And the images of nature that scientists share through this cooperation are the resources they hold in common and apart from the rest of the world. They are what help keep scientific knowledge more the property of scientists than it ever could be for nonscientists.

Scientists not only use one another's work to develop their tacit understandings of natural process; they also play off of scientists' tacit knowledge to try to gain power in intellectual debate. They attack the tacit knowledge of their fields and make problems visible when they had been "hidden" before. At least they try to do that.

For years I've gone around and every time I go into a public lavatory, I smell that stuff which is paradichlorobenzine, a little cake of stinking stuff hanging around. So you smell it, you know, and somewhere along the line you read its label and it says paradichlorobenzine, and that stinks. Later you're reading through something and you find that paradichlorobenzine is a mytotic poison. It's the thing that interferes with cell division. Now, why do we want to go around breathing that? You know, I mean you're going to fiddle around with things you breathe, why, let's just stay away from those very subtle poisons like mytotic ones. So then you begin to wonder, well, what happens to these cakes? Do they evaporate?

They go down the sewers and they're discharged into the sea. Is that good? Anyway, I started in, chewing away to get somebody to analyze this stuff. Very difficult analysis, or at least they thought so because it would be there in trace amounts. Well, it isn't there in trace levels at all; it is the primary chlorinated organic material that's going out in the waste waters along the West Coast. [Interview with a senior scientist at a major research institution]

Issues that have remained unexamined in the past may be opened up for serious study this way. People will start to chart the movements of substances like paradichlorobenzine. Procedures (like the use of labeled CO_2 mentioned in an earlier quote) that have been standard practice in the past will suddenly come under criticism, and new techniques will be introduced to replace them. The result is a new measuring system that allows study of aspects of nature that were simply not approached before. Where these shifts are successful (meaning that others follow the lead), the result is conceptual shifts in the field with ramifications broader than merely procedural ones. Scientists try to get colleagues to "see" nature in a new light and reassess old knowledge in the light of the new.[20] In the following case, a research team made a well-known phenomenon problematic, explaining it in a new way that raised larger theoretical questions:

We call them overlapping spreading centers. Instead of spreading centers that are offset by transform faults in the straight, orthogonal(?) way, the spreading centers overlap. And that sort of . . . that in itself is not that important. It's an interesting new way that spreading centers behave, but it sort of represented a new structure which required explanation, and in the process of trying to explain it, it started to take us to sort of new ways of thinking about . . . the way the seafloor was formed. So it sort of opened up the door that had been closed. And so the law's interesting. [Interview with a senior scientist at a major university]

Changing the domain of science in this case and others is one strategy scientists use when they vie for power in their fields and subfields. To the extent that they find something that opens up many new avenues for research and requires some rethinking of existing work, they make their work powerful in shaping scientific debates and practice. They serve science by serving themselves this way, making claims for themselves that can be used to make claims for the institution of science itself.[21]

Nonscientists who enter into the world of research show many lapses in their knowledge of science. Even when nonscientists become knowledgeable about one area of research, their lack of education in the scientific specialties makes them make mistakes that no scientist would commit. They lack the fundamental tacit knowledge with which scientists think.

Good evidence of this lies in the tapes made on the *Alvin* submersible during a dive to the hydrothermal vents made by Walter Cronkite and a marine biologist. On the tapes, you can easily tell that Cronkite has read a great deal of the vent literature to prepare for the dive; he knows major features of vent sites to look for, but he does not know how to distinguish important from unimportant features in the objects he observes. He does not know enough general deep ocean science to distinguish between those observations that are scientifically important and those that are not.

PILOT: Hello, *Alvin* at the bottom. See clams out the front window and a tall chimney and some vents. The depth is 2595, 2-5-9-5, Navy pressure 1556, 1-5-5-6 at 1 decimal 5 degrees. Outside temperature 1 point 7 3, 1 point 7 4. I might be having trouble, Bob, with the propulsion. If I go down, then I can't move the rudder or can't operate the starboard arm but hit-and-miss. I'll let you know if I uncover anything else. Over.

CRONKITE: Ooo, ahh. [Stretching]

RADIO VOICE: [Scrambled]

PILOT: We don't have any rudder.

CRONKITE: *We don't have any rudder?* [Alarmed]

[PAUSE]

CRONKITE: That looks like a big fish over here.

OTHER: Could it be?

BIOLOGIST: Has it got a long, tapered tail?

CRONKITE: Yeah.

BIOLOGIST: It's probably a rat, what we call rattail. They're very common in most deep sea areas, not just associated with vents.

CRONKITE: We're settling down to the bottom. Is that what we mean to do?

BIOLOGIST: Yep.

CRONKITE: Got a big rock over here.

PILOT: Oh, no problem.

In this case, Cronkite does not know how common the rattail fish is and whether the problem with the submarine's rudder is serious. He finds a common fish very interesting, and he is very anxious about a rudder problem that is apparently not bothering anyone else. Throughout his dive he makes similar "mistakes" in evaluating his perceptions.

He also betrays a naiveté about science when he expects his co-diver (a biologist) to be interested in geological formations, while the biologist is trying to point his attention to the biological feature of the vent sites that has gotten the most press attention: the tube worms.

> CRONKITE: Here's a perfectly round, a magnificently symmetrical round basalt deposit. Absolutely a perfect dome.
> BIOLOGIST: Do you see some of the clam shells all over?
> CRONKITE: Yeah. Are they the white things?
> BIOLOGIST: Yes.
> CRONKITE: Yeah.
> BIOLOGIST: Okay, Walter, if you want to see something spectacular, can you make it to my window?
> CRONKITE: Yeah.
> BIOLOGIST: You could try. There you go. If I can get you guys to shoot that . . . color video out there [inaudible].
> CRONKITE: Oh yeah! My God! Ooo. What are those things that look like bones laid up in there? That's tube worms?!

Even when Cronkite is showing off his education about the vent fields, he still displays a kind of conceptual awkwardness. This makes his conversation with the biologist sometimes strained, while the latter tries to figure out what bit of information Cronkite has pulled out of context to organize his perceptions. What continues to characterize Cronkite's observations above all else are the limits of his knowledge about the rest of the ocean that would allow him to evaluate with better effect the importance of the objects he sees at the vents.

> CRONKITE: Huge deposits of these clams, aren't they?
> BIOLOGIST: Yeah.
> CRONKITE: Hordes of them within a few feet. This is not the way small clams, clams in the rest of the ocean, look like, is it? They don't live in great piles like that, do they?
> BIOLOGIST: Well, the things look different dead, but . . . well, even when they're alive they look a lot like that. And you're right, the shallow water ones, no. You don't see them piled up like this. But we feel that these animals—like I was trying to explain to you about the tube worms, too—these animals have the chemosynthetic bacteria associated with them also. It seems like maybe they do not have to be like a filter feeder, get all their nourishment out of the water. They might have the symbiotic bacteria in them, so they don't have to be separated from each other, as individuals.

CRONKITE: There's something going by every once in a while, like small shrimp of some kind.

BIOLOGIST: Yes. Those are . . . they're kind of reddish brown?

CRONKITE: Yeah.

BIOLOGIST: Yeah, those are not specifically associated with vent systems. You find those in other deep sea areas. They're a . . . what we call a decapod, much like normal shrimp that you eat. They belong to the same group of organisms. But they're not peculiar to the vent systems. You'll see them.

I draw attention to Cronkite's difficulties in interpreting what he sees, even in the light of what he had read, not only to indicate how much background knowledge scientists bring to even such a new finding as the vents at the point that this tape was made, but also to suggest how difficult it is even for a conscientious student of science who is not a scientist (such as Cronkite) to be able to interpret and use scientific literature for his purposes. Cronkite's problems can be taken as models for the problems faced by agents of the state trying to put science to use. Cronkite is better off having a scientist with him to interpret what he is seeing rather than relying entirely on written material, and similarly agents of the state are better off, in trying to apply science to policy issues, asking scientists for advice rather than reading scientific documents.

The reason for this is quite clear from these tapes of the Cronkite dive. Cronkite may know a lot about the vents, but he does not know what he does not know, and he has no way to estimate how much scientists really know without asking them. New analyses are made all the time, and no one without a stake in following a field in detail is in a position to assess them. That is why consultants more than funding reports are the major link between science and the state.

That is also why scientists have more to learn from the results of scientific research than most of the funding sources. Scientists have a better sense of the significance of new findings beyond those claimed by researchers, and they have better ideas about the weaknesses of results. The world of ideas, however much it may be coveted by nonscientists and made possible through funding, is in the end a world of and for scientists. And it is scientific expertise, knowledge, and skills embodied in scientists that are cultivated through state funding. Scientific advisors are the ones who can use the voice of science authoritatively (both in spite and because of scientific disputes), and bring it to bear on policymaking as a political tool for the modern state.

CHAPTER TEN

The Voice of Science[1]

THE COMBINATION of dependence on and autonomy from the government among soft-money scientists (resulting from their fiscal dependence and drive for intellectual control of the natural world), combined with the use of scientists as advisors by the government, have turned scientific researchers in the United States into an elite reserve labor force. Because scientists have been called upon by government to consult on issues of political weight, they have been encouraged to conceive of themselves as influential and substantial experts. As a group, they have developed a social image as an elite with a powerful understanding of the natural world and (for some) special access to power (or the powerful). At the same time, they have made, or been given bases for, claims of elite status because of their relationship to government, but they have also been influenced in their studies by the practical interests of the state in science. They have been trained to acquire sets of scientific skills with use to the government for its practical applications in technical—often military—innovation, for shaping and evaluating policies, or for defining the cutting edge in scientific research eliminating and out-moded and discredited lines of research.[1]

The fact that scientists seek autonomy in the funding relationship, resisting efforts on the part of the government to shape the direction of scientific innovation, has been blessed by the state at the same time that practical concerns have shaped funding policies in government agencies. This concession to scientific autonomy had been more than mere lip service. It has been fundamental to the use of science for government action. Science gains value to the state because of its claims to "independence" and "detachment." The voice of science is authoritative to the extent that it seems objective and above politics even when applied to policy. Scientists use their degrees of institutionalized autonomy as evidence of their elite status and intellectual independence, but ironically, the government profits from the relative autonomy of scientists because scientists' belief in their own detachment helps give the voice of science the legitimating power that makes science a resource for the state.

As we have seen already, the institutionalized autonomy of scientists is much less than scientists (and many students of science) think. Peer review may be a precious power to elite scientists, but it is not really a source of autonomy for most researchers. Designing their own research proposals may also be a valued right to scientists, but it is hardly the same thing as being able to set one's own research agenda, when so many proposed projects are not funded. And even researchers who routinely find funding for their projects can find their autonomy circumscribed by unsuspected external forces. The methods they take as legitimate routes to knowledge may have gained their importance in the scientific community because people in government supported so many applied projects requiring these techniques that the methods became sophisticated. This is how sonar studies gained their important role in marine geology. Similarly, the scientific problems that interest researchers may seem to respond only to the work of other scientists; but if other scientists shape their research agendas to fit the funding priorities of the state, the scientific problems interesting to researchers may be deeply structured by government needs. In addition, scientists may experience (without noticing it) technological limitations placed on their research by the politics of the military, their contractors, and businesses that manufacture research equipment. Scientists may decide that some research dreams cannot be realized because they are not yet technically feasible, but in fact the technology they wish for may already be possible or extant. They just cannot know because they are not given access to this knowledge or to this equipment.

Belief in the power and autonomy of American scientists may be politically valuable for the state, but it is also a historical artifact. Many senior scientists living today still remember the demobilization of science at the end of the war, and can remember and convey to their junior colleagues how dramatically different research is today from what it was during wartime. Researchers who began serious research careers only after the war may not remember this, but they were raised (so to speak) in the cold war period, when a great deal of political debate was organized around the advantages and disadvantages of thinking in an "open" society. The work of American artists as well as scientists was taken as evidence of the creativity unleashed by the intellectual autonomy of "free world" thinkers (as opposed to the "controlled" intellectuals of the Soviet Union). For overtly ideological purposes, then, as well as for practical ones, American scientists of the 1950s and early 1960s were taught that they had a great deal of autonomy in their research and that it was very good for them. There was (as we have seen) some institutionalization of this ideology. Early Office of Naval Research funding was given in block grants without much oversight from the military; but even so, not everyone was funded. When the block-grant system was dropped, it was replaced by the proposal-based funding system that lessened the discretion of individual scientists. The peer review system instituted at the National Science Foundation was meant to provide some new autonomy for scientists, but it is not as great a source of autonomy (even for elite scientists) as it was assumed, in any case. The administrative power in the system has simply been too great.

There remains an important legacy, however, to the postwar period's funding system that defines some real autonomy for science: the right of scientists to use even research funded for applied purposes to address issues of interest only to the scientific community. Where research is not classified or otherwise controlled for overtly political reasons, scientists are really *expected* to make scientific sense of their research results and are often encouraged to develop research designs for their projects that will facilitate this process. The reasons for this are simple enough. To the extent that they keep their scientific reputations good, they increase the credibility of their expert opinions, and to the extent that they do good science, they bring greater expertise to applied programs. So pursuing basic science has served rather then undermined the state, as long as it has been compatible with practical problem solving.[2]

Perhaps more importantly, the world of scientific discourse has remained primarily in the hands of scientists. The government may limit the subjects scientists study and their methods for research and analysis, but they do not control the ways scientists use their research materials to address the world of science. The world of science has its own political dynamics that make the status of any given piece of research problematic to read for an outsider. Scientific papers, for all their logocentrism, can have multiple interpretations. These are most visible during periods of intense intellectual debate (what Collins[3] calls "extraordinary periods"), when competitors tear apart each others' work by reinterpreting their data.[4] During these times alternative readings of research results are part of the public life of science. During other periods, scientists often read papers in exactly the same way (just less visibly), looking to see whether their understanding of others' research fits with the published explanations. As we have already seen, when reviewers for journals read papers, they are often absorbed with how the work impacts on their own, and thus they provide a very personal reading of the findings. Even when they have fewer stakes in their readings, scientists often prefer to begin reading papers by studying tables and other data displays rather than the text. They want to see how they might read the research result without being influenced by the text (which functions as an extensive caption for the images).[5]

The indeterminacy of new findings, their political meanings, and their contested status all make scientific literature a mine field for outsiders who want to use their results to delineate policies. The truth is that they often do not have "results" in the sense that government policymakers would like. They have methods for interacting with nature and for making interpretations of what they have done. The meanings of research projects are not clearly defined. For this reason, what scientists "know" is a construct—something that scientists make up from the wealth of data and publications that come to their attention, a synthetic image of the natural world produced from the activities of an intellectual world. The image of nature created this way may be synthetic, but it is important. It makes it possible for researchers to act and feel as though they know something about nature that people outside of science cannot know.

There are others who make a living trying to keep on top of developments in science. Nonrotating program officers working in funding agencies, for example, take it as their job to know what is happening in science, but, as we saw already, scientists who advise them or join them temporarily as rotators try to shape funding patterns by selectively tauting and demeaning different lines of research. So what agency personnel know is partially obscured by the politics of funding.

For the most part, then, the world of scientific knowledge is left to scientists and is closely guarded by them.[6] Mediators or translators from the scientific community speak for science to the state, using specially designed representations of scientific research that are cleaned of many of the technical issues that are central to the activities of researchers but not important to outsiders. In this way these translators produce what researchers in the sociology of science call "boundary objects," images of the natural world based on science but not so highly codified that outsiders cannot understand and use them.[7] These people and their versions of science constitute the voice of science that is given to the state.

Not all scientists take on this role as translator. There seem to be three types who do: elite scientists with an interest in policy (government policy, science policy, or both); elite scientists who do applied work as one part of the work of their labs; and nonelite researchers (often in someone else's lab) engaged in applied projects.

The first type functions as a science advisor of sorts, and institutionally named science advisors are drawn from these ranks.[8] A subgroup of these policy-oriented scientists is most concerned with the relationship between the state and basic research. They frequently work with NSF and other funding agencies, agree to serve on National Academy of Science task groups studying applied problems, and perhaps write on science policy from time to time. Another subgroup works on weapons development and strategic policy. They are less concerned with NSF and NAS than the military and its relationship with science, and the usefulness of science for strategic planning. A few people will do both and become the major translators of science to the state. They use an array of contacts within the scientific community to keep track of developments in science and their potential consequences for policy.[9]

The vast majority of those who do applied research are not major mediators between science and the state. They do not represent science as a whole to the state, and they do not address basic issues of policy like official science advisors do, yet they engage in similar activities. They too use their networks to learn what is happening in their fields, and they too make their expertise available to the state. The difference is that they translate only small areas of science for the government at any given time, those relevant to the applied task at hand.[10] Hence they do not so much define science to the state but show how to use the skills and language of science in one small area of policy. As such, they may engage in a less dramatic and public mediating role, but it may in fact be socially more consequential an activity; producing for the state a language of science and a practical version of science with which to further political action.

By acting in the world of taken-for-granted culture, these scientists do what they take to be quite mundane and perhaps alienated work, engaging in activities that seem banal to them and remain for the most part invisible to the rest of us. But there is enormous power in what they do— in part because it is taken for granted and not contested publicly like more visible areas of policy. They keep a pact with the state by giving counsel, but they do it in day-to-day routine meetings that do not reach public attention. In these meetings we can see best the dependence of scientists on the state and the price they pay for it. There is no contest for autonomy here. The voice of science is presented to the state for political use. This voice is not the same one heard within the scientific community. It is a public voice, a boundary object that can make sense both in the world of science and in the world of policy. It is a voice, like all voices, born out of an interaction between two parties.

Scientists do not always pay much attention to the advice they give. They do not give away anything that they need or value (such as ideas for a scientific article). The advice they provide to bureaucrats is, in their minds, simply a less fine-grained interpretation of scientific studies, just a less good version of the analyses they make every day when they interpret and reinterpret each other's work. The interpretations they make for the state are usually not ones they are particularly interested in having associated with their names. Most of them prefer to think of themselves as detached thinkers, formulating models of nature, than as advisors on policy issues such as the disposal of nuclear wastes. Scientists doing applied work are required by their funding agreements to provide the advice they give, but they are also trained to feel appropriately reluctant when called upon to do so. Even when they discuss policies, scientists often consider themselves not very useful advisors because they feel they must cover any policy recommendations they make with provisos to make them scientifically precise. They know that their attempts at precision frustrate the directors of their research programs, but there is little they can do about it, other than be a bit embarrassed. They are trained, after all, to think like scientists, and they do. But the sense of frustration with scientists that they see in government-agency personnel makes them think less of the advice they provide. They think it has relatively little value to their state patrons. But the consequences of their actions are much more significant than they might think. For the voice of science—with its authority and dispassion—has been a central political tool since World War II. It has been indispensable in the development of weapons and strategic policy; it has shaped waste disposal policies, social programs, methods for handling (or not) environmental problems like acid rain, and government support (or not) for medical care; and every year it is used to articulate fishing regulations, the use of pesticides, and the drugs that will be available on the American market for doctors to prescribe and consumers to use. The elite reserve force of scientists contains a wide array of experts to provide scientific rationales for myriad small government decisions, and this vast number of cooperative scientists gives to politicians a powerful resource for legitimating their policies: the voice of science.

This is not to say that when scientists lend their voice to a program, it will necessarily prevail or even that the scientific opinions imparted to the government are carefully enough formulated to be unassailable. Scientific opinions may be designed to be fortresslike, but they are also open to assault. Moreover, science is so fraught with controversies that it is usually quite easy to find an articulate and respected researcher with a view at odds with the position presented by another scientist. But the power for politics of the scientific voice does not lie in the truth of what scientists say or even the value of their truth-claims. What matters is that the opinions of those who legitimate government policies cannot be disregarded. Everyone may not agree with them, but if politicians opposed to a project cannot show that the voice of science used in its defense is fundamentally flawed, if it is just a matter of opinion, then there is no way to use science against the project, and the science used to support it will likely prevail.

Take the example of the Strategic Defense Initiative (or SDI, or Star Wars) weapons program or defense system, as the case may be. This is a highly controversial program even in the scientific community. Its controversial political status has certainly been matched by the degree of skepticism about its feasibility. Why did the policy survive despite the barrage of criticisms? Because the credentials of those who supported the program are first class. They are not flakes or light-weight scientists. They may be politically conservative or longtime advocates of the policy of nuclear deterrence, but that does not affect the power of the voice of science they wield. Others may doubt the wisdom of their counsel, but they cannot completely undermine their position. The fact that science always contains elements of controversy becomes politically useful rather than troublesome in this situation. It is easy enough for members of an administration promoting the program to argue that scientists never agree, so you cannot wait for unanimity before acting. The fact that some substantial and authoritative scientists support the program is enough justification for pursuing it. Money is made available for research relevant to the program, and scientists line up for it. Only a change in administrative programs will alter the policy.[11]

Because the state holds the purse strings for the research establishment in the United States, it has enormous power in articulating and legitimating policies. On a routine basis, people in government are able to have access to the voice of science for policymaking—for shaping policy, legitimating it, or anticipating controversies. Opponents of government policy often do not have access to this voice to support their own positions.[12] Where the stakes of the controversy are great enough to mobilize scientists politically, it is possible to find scientifically couched arguments against policy, as there have been against the SDI program. But often opponents of policy are supported in their own research by the very institution whose policies they abhor, and many scientists are simply reluctant to make trouble for themselves by speaking out unless they feel so strongly or are part of a large enough movement that they think they can, as well as must, speak.[13]

In spite of their vulnerable political position, scientists have profited enough from the relationship of science to the state in the United States, harvesting so many intellectual riches from the research that the government has made possible, that they usually do not resent their situation (at least on these grounds). They feel that they have the right to speak out if it is necessary, so they do not feel imprisoned in silence and held completely outside the public realm. Still, scientists who are specifically recruited for applied research on a particular project are often asked to keep their opinions confidential and give them only to their funding contact in the supporting agency. They seem to be powerful enough, since they give their advice as *insiders* to government. But they do not often realize the price: they must be effective in shaping policy or suffer the results in silence. Being asked for advice at the point of policy formation seems an opportunity for the exercise of power, but it is often a moment in which scientists take vows of silence.[14] Still, in a democracy people are used to losing in political debates, and scientists can justify their losses in these terms.

What reassures scientists the most when they face the power of the voice of science and their powerlessness to use the voice in the public arena is the idea of their autonomy. Scientists are not, in the end, politicians, and they suffer political defeats better than loss of face among their peers. As long as they can conduct research with which they can advance their science (both science itself and their positions in it), they can feel potent. But the cost is that scientists cultivate an expertise that empowers someone else.

I am not trying to imply here that scientists would make better policy than politicians and ought to be given more power. I do not dream of scientist-kings as perfect rulers any more than I dream of philosopher-kings or any other saviors. The point is only that what scientists give the government is very powerful, and what power scientists exercise is minimal. Instead of power, they have been given tokens of autonomy, a phantom power for holding the government at bay, for keeping science pure, while the state has as much interest in this "purity" as the scientists themselves.

This system is held together only because scientists can profit from the work they do for the state. The same piece of research can yield results embodied in two ways: in publications and in the expertise or skills of scientists. Science embodied in texts is used to empower science and hold in check the power of the state to direct science. Science embodied in the scientists themselves and in their counsel is expertise that the government can call upon for formulating policies and legitimating political choices.

Take the example of some scientists who studied the vertical movement of lifeforms in the ocean to counsel the Department of Energy on the feasibility of deep ocean disposal of nuclear wastes. The government wanted to know if radioactivity leaking on the seafloor was likely to affect the fish caught by commercial fishermen, so they were interested in things like the vertical mobility of organisms. Earlier scientists had paid more attention to the relationships among the animals living at any given level of the ocean than vertical mobility, so scientists were interested in the problem for their own purposes. They did not experience much loss of autonomy in these mobility studies, even when they were doing it to act as policy advisors. One scientist I talked to said that one always felt some loss when doing applied work, but not much in this program. This was because they could successfully serve science with the projects funded under this program.

At the same time that these researchers working in this waste-disposal program were giving their advice to the DOE, another colleague, writing about the results of his study of population ecology, took the position that he could and should not really make policy with his research results. He delegated this responsibility to politicians. What did this mean? It was not that he had no policy thoughts of his own. He had plenty. It was not that he did not want to tell anyone his ideas. He routinely told people in the DOE program what he thought. What he meant was that he wanted to be *seen as* a scientist, not a bureaucrat from the government, and that meant *to him* that he should not attend to political issues.

To the extent that this man and other scientists in the program were heard on the subject of waste disposal, they were heard by government bureaucrats or provided lines for the bureaucrats to use. Some took a more active political role when they agreed to speak as government experts at public meetings; but in doing this, they called into doubt their scientific credentials, and made themselves seem more like government functionaries. In either case, they were made the voice of the state or had no voice at all.

The process of giving the voice of science to the state for its political ends is in formal terms the opposite of ventriloquism. Scientists do not send their voices out to speak through the mouths of mute government officials. Government officials extort the language of science and scientists' analytic skills to do their political jobs. Scientists are made mostly mute, except when politicians find their voices useful.[15]

Ironically, most American scientists find it an honor to advise the government by serving in the National Academy of Science, testifying before Congress, or being part of a commission studying some problem or another. The fact that the government wants to use their voice seems very grand. And they have some grounds for seeing honor in it. After all, a scientist's voice can have some *power* in the mouths of others. For the most part, researchers *only* have public influence when politicians champion and promote policies with their advice. So to be heard politically, they have to arm politicians with their expertise and the legitimacy of the voice of science, meanwhile keeping themselves more or less mute as scientists.

Surprisingly, the act of giving advice helps to produce univocal voice of science that is not available for the use of individual scientists. In the situations where scientists function as consultants, they are often under constraints to reach some consensus on policy recommendations. But researchers speaking their own minds as individuals do not have the right to speak for "science" in the same way. So collectively scientists give to the state a political tool that they cannot use very easily for themselves.

The dual use of science for science and for the state serves to enhance the authority of scientific thinking, but it also gives its power to politicians. Since the opinions of scientists are often presented as *above poli-*

tics (and couched in the language of dispassion), politicians can use them as indisputable support for some political position. Scientists feed this pattern as they struggle for their own purposes to defend their objectivity, but in practical matters they are less dispassionate than mute.

Jürgen Habermas and Alvin Gouldner[16] have argued that the politics of communication can change depending on whether information is communicated in oral or in written form, and whether messages are filled with information detached from a speaker or they are expert opinions expressed by a person. Bruno Latour[17] has argued that written words and drawn pictures constitute "immutable mobiles," pieces of paper that can be carried over long distances without changing. Thus they do something quite different from face-to-face communication; they extend the power of language as a socially structuring device over larger geographical and cognitive spaces. These authors, each in their own way, suggest that the social consequences of words have a great deal to do with their object-life. Certainly this seems to be the case with science.

Science is controlled and communicated in two different ways with opposite effects on the power of science. Science-in-texts controls and communicates scientific findings and ideas within the social world of science, and helps empower science as an institution. It is a world of communication-in-code that remains for the most part within the social world of science. Science-as-counsel, carried in oral discourse, is put, in part, at the service of the state in exchange for funds. It is communicated verbally, for the most, and with it, the voice of scientific authority is given to the state for policy objectives. Thus it empowers the state. In these patterns, we can see how the different systems for communication and control of science are used to serve different systems of power.

Since it is obvious that scientific advisors routinely write down their recommendations and that scientists doing basic research in the laboratory routinely talk to one another as they try to make sense of their research results, let me say a bit more about what I mean by oral and written discourse in scientific communication and the communication of scientific expertise to the state. I must begin by discussing a bit more the nature of oral discourse in literate social groups.

The advice given the government by scientific "experts" gains its authority in part because it is based on the highly literate activities of scientists in their labs. One cannot legitimately talk about scientists as members of an oral culture, even though they often work in groups and spend inordinately long periods talking to one another about their experiments and their results. But what distinguishes science and has helped to make it grow in the West is the use of documents—records of experiments, expeditions, and other studies—and reflection on these records that result in written journal articles and other papers. The science used to formulate policy is developed from written words, numbers, graphs, and pictures registered on paper in scientific labs. They spew from image analyzers, chemical analyzers, word processors, and other kinds of computers and research equipment. They are born of the technician/machine symbiosis described earlier in this book. These traces are made meaningful and given their value to science by being placed in the contexts of scientific debates, and these debates are formally located in academic journals. This is not to say that scientists do not talk to one another and argue with each other at conferences and colloquia. The public arena in science is primarily in the world of print, but it has an oral element as well. Still, most of the thinking done by scientists is organized around pieces of paper. It is a deeply literate activity.[18]

Scientists bring to the government the fruits of a highly literate culture when they provide counsel, and they also bring advantages of oral discourse. They may come to meetings with position papers that they present by reading, but their counsel is oral in many ways. Counsel is usually offered in group meetings and evolves from the discourse among those attending the meetings. Scientific experts are often also scholars of a rather narrow range of intellectual work. Policy proposals do not, in most cases, fit nicely into the specialities of scientific researchers, so discussions of policy are often set up to bring together a variety of specialists with knowledge of a problem. The advice given the government is created during these meetings through the interaction of participants. It reflects less the nature of the position papers than the conversations forged from them. It is a product of oral discourse, but not an oral culture. It emerges from oral discourse situated firmly in a literate culture.

Advice is also often contingent. Scientific knowledge is meant to be absolute, and the fact that it is meant to be etched firmly in cultural memory (on paper) expresses that ideal. Advice is not that way. Scientists giving advice are frequently faced with thinking about issues that are not usually at the forefront of their reflection. The advice is solicited in response to government or social problems and therefore must be contextualized in social worlds not always so familiar to advisors. So they think contingently: if this, then that. Contexts for making policy decisions are developed in the process of meetings to make recommendations, and responses are contingent on assumptions about the problems that emerge from the meetings where they are discussed.

The results of these meetings are often less than clear-cut, in spite of pressures to reach consensus. The situations are filled with uncertainty, and the results reflect it. In the absence of unambiguous endorsements of some policy position, one of the most valuable contributions scientists make to the discussions is the delineation of a language for addressing the problem. They develop ways of *talking about* public issues. It is literally the language of science, then, more than the ideas of scientists that is made available for public debate. Of course, there is not one single voice of science unveiled at these meetings. Different voices are placed in competition with one another to define the problems and solutions. Part of what is conveyed in these meetings is what scientific voices win debates over others. The power of competing scientific rhetorics on public issues can be assessed from these meetings and used in formulating political strategies.

Written reports might be able to convey some of the power of different rhetorical styles or approaches in science, but oral debates are more efficient means for testing and comparing them. Moreover, locating these debates to a private sphere incorporated within the institutions of government gives the government greater control over the use of the rhetoric. What is said by the scientist that seems to retard a desired policy can be kept confidential *because* it is so effective, just as it can be used publicly when its rhetorical power seems to fit the interests of the bureaucrats who contracted to hear it. Oral discourse is easier to control this way. It cannot be spread independent of the persons who originally thought it or listened to it. Perhaps more practically, what is said in oral discourse can be forgotten or misremembered, so it is easy to deny.

I call what is given to the state at these meetings the voice of science. It is a voice of many senses of that term. It is something spoken more than written, and it conveys the rhetorical style of scientific thinking and argumentation. The value of it lies, of course, not so much in its orality or its rhetoric; it lies in the valorizing character of science. The image of science as a means for achieving lasting knowledge, detached analysis, and thoughtful reflection gives this voice a power that makes it politically useful.

Scholars of the nuclear arms debate have made this point. The fact that the forces supporting deterrence in the nuclear arms debate have had the rhetoric of science on their side has given them enormous power.[19] Opponents have been accused of being all the things scientists are not supposed to be: ignorant of the facts, too emotional to be detached, and quick to act rather than thoughtful. Even when armed with the same facts, the forces of deterrence have seemed more powerful than those supporting disarmament because their justifications of their policies have been couched in the socially recognized codes of scientific thinking. And even when scientists have tried to use their rhetorical training to oppose deterrence, they have been accused of not knowing the facts (because opponents of the policy have not been given access to the same set of facts or at least cannot know that they share the same facts). So between government control of expert estimates of the Soviet threat and the rhe-

torical power of science, the government has been able to support its
policies and discredit the forces supporting disarmament.

The result of this situation is ironic as well as dangerous. There is an
apparent growth in the power of science in American society, as government relies increasingly on scientific rationales and programs for policy-
formation and realization, but it does not create that much real power for
scientists. The power of science has become part of the power of the
modern state because it gains power in the hands of policymakers. The
power of science today resulted in large part from what governments
asked scientists to do during the Second World War. The success of scientists in serving government then gave them reason to ask for and receive ongoing government support. But what they did during the war
was not so much to empower science by advancing it intellectually; it was
to empower the state (the military). The intellectual benefits to science
came mostly later in the attempts to think more precisely about what had
been done during the war, and to work more extensively with the new
technologies that had furthered wartime research.

Since that time, the situation has not been radically changed, although
the legitimacy of scientific funding has been questioned from time to
time and budgets have shrunk and expanded again. The power of science
is still measured more by its practical applications and policymaking potential than by its abstract elaboration. Soft-money scientists in the meanwhile have not gained that much power. Their relationship to the government has given enormous wealth to research institutions and
universities, giving them the capacity to expand their physical as well as
intellectual domains. Still, individual researchers have been faced with
anxious days waiting for letters from NSF or ONR.

The power of science is not even so much in the machines scientists
help produce for the military, the weapons systems they advocate and
design. It lies in the voice of science that goes with them, the authority,
the sense of hope and power. And the power of this rhetoric reveals less
the institutionalized control of science in modern societies than the success of the state in using the rhetoric and its applications to advance its
interests.

Scientists are more vulnerable. They are an elite, without any doubt,
but they are also a reserve labor force without any mission of their own
in the public sphere except to make sure that the research budgets get
bigger. In many cases they give up their rights as citizens to debate *publicly* the virtues of policy. As advisors, they agree to having their voices
expropriated, and they do not think much about it. They are taught to
equate their power with their autonomy from the state, the lack of direct
state control of research. But that is no measure of power when the state
needs their detachment to serve its interests. They are left the world of

science to govern internally, just as nineteenth-century women were given the domestic sphere to run, and they run it with the same mixed effects. It is, after all, an independent world to relish and flourish in. But structural dependence is a source of constraint. The quid pro quo can be costly. And in the relationship between science and the state in the United States, the scientific establishment routinely gives away one of its greatest assets: its voice.

Notes

CHAPTER ONE
SCIENTISTS AS AN ELITE RESERVE LABOR FORCE

1. The head scientists were not in much trouble; the Principal Investigators had always been supported at one-third time or less, so they were still getting salary from elsewhere. But some of the junior scientists in the laboratories had received their salaries entirely from the Department of Energy; overnight, a number of them were no longer getting any income. In some labs, they reduced their work load so that they could maintain one-fourth salary for a longer period, while they looked for other work.

2. The withdrawal of funds from these scientists is interesting in part because it is so perplexing. These people were team players, with high prestige, who at least in theory should have been less vulnerable to government dismissal. They had looked at an important policy issue as experts wielding lofty theories, and they had received high respect from DOE officials while they were doing the research. They also served the DOE well; they provided, for the most part, encouraging counsel, supporting (by and large) seabed disposal of wastes. Their studies showed ways that they thought it could be safe—at least a lot safer than disposal underground near sources of drinking water. Some had even testified on behalf of the military when the Navy wanted to scrap old nuclear subs by sinking them in the Pacific off the California coast.

Perhaps they had lost some of their usefulness and prestige in the eyes of the government because they had lost this argument to the locals who were militant in their opposition to the dumping (many being fisherman who were worried about their livelihoods). Perhaps the political problems surrounding plans for seabed disposal of wastes helped to spur the decision to forget about it, at least for the present. In that case, one could argue that these scientists lost their funding because they lost a political battle, and this made them lose their political clout. There is some truth to this, but this analysis hides more than it reveals. It ignores the cultural power of science, and only begins to approach the question of why the government provides money for scientific research and how scientists negotiate the opportunities and vulnerabilities built into this fiscal dependency. For some discussion of the continuing politics of land-based disposal, see Crawford 1985a, b.

The larger issue of how professionals gain power through control of information is discussed in a very useful fashion by Freidson 1986. Freidson makes the point that knowledge does not necessarily empower professionals. To understand how it does or does not give them power, Freidson argues, one must look at the practices of the professionals, the way they mobilize the knowledge. Structural studies alone will not illuminate the sources and limits of professional power.

3. For a discussion of the cultural power of science and its historical develop-

ment, see Starr 1982. See also Dickson 1984; Greenberg 1967; Lakoff 1966; Shapley and Roy 1985; and Sherry 1977.

One would think that the social position of science vis-à-vis the state would be such a central sociological issue that there would be a major body of literature on the subject. But because the sociology of science has developed as a subfield of the sociology of knowledge, this is not the case. John Ziman mentions it in Ziman 1984. So does Richter 1980 and Gummett 1980, but pieces on U.S. science policy are not so common. See Price 1965 and the classic, Dupree 1986. More recently, see Cole 1983. This book is particularly intriguing because it raises the question of when and whether government can and should enter into restraint of scientific research. The author argues against the contemporary assumption that American scientists have, want, and deserve autonomy. While I disagree with his conclusions about scientific restraint, I think this questioning of scientific autonomy is both serious and very important. Another unusual and important model comes from Gusfield 1981. Gusfield discusses the uses of scientific language for policymaking. More than anyone else, he documents what I will describe in this book as the use of the voice of science for political purposes.

Most sociologists do not share this point of view because they have seen a higher calling in using the sociology of science to revitalize the sociology of knowledge, not in addressing policy issues. They have been concerned with using social variables to explain scientific ideas or the content of scientific texts. See, for example, Mulkay 1979; Barry Barnes 1977; Merton 1973; Mulkay 1972; Merton 1970; Feuer 1963. One might still think that the peculiar structural position of scientists would have gained some attention, given the fact that in Mannheimian terms the structural location of scientists ought to be central to explaining their scientific ideas. But there seems to have been a division of intellectual labor in the field that encouraged historical sociologists of science to study the institution of science and its fate, and they left students of modern science to explain the ideas of scientists from dynamics within their social worlds. This has encouraged the bulk of sociologists of science to study the social world of scientists more than the social role of science. Laboratory studies (internalist analyses) emphasize the production of texts; see Latour and Woolgar 1979; Latour 1984, 1986; Knorr-Cetina 1981; Knorr-Cetina and Mulkay 1983; Gilbert and Mulkay 1984; Law 1986; Lynch 1985. Citation and other prestige studies emphasize patterns of education, research, and publication as indicators of success and their relationship to the development of ideas. See Cole and Cole 1973; Zuckerman 1977. Historical research on science tends to emphasize the impact of texts as carriers of conceptual schemes or the power of scientists in crafting conceptual systems; see Kuhn 1970. Holton 1973; Merton 1970, 1973; Sarton 1957; Barnes and Shapin 1979. In all these cases, what is important is the production of "science," be it embodied in or equated with publications or not. The communication of information is at the heart of the scientific enterprise, and production of information for communication is at the heart of research practice. The one major sociologist who has recently tried to explain the growing power of science is Paul Starr. But his book *The Social Transformation of American Medicine*, was devoted to explaining the power of the American medical establishment rather than research science per se. Happily, he lingers on the development of the medical research establishment and its importance to the empowerment of medicine in the twentieth century, but the book does not (and was not intended to) provide an adequate account of the power and limits of nonmedical science in the United States.

Journalists and political insiders to the world of science often comment on its social role. See Dickson 1984; Greenberg 1967; Lakoff 1966; Shapley and Roy 1985; and Sherry 1977; but they have not taken a close enough look at the organization and consequences of the soft-money funding system to notice the differential use of scientific information and skills by the state. They have contributed to the view of scientists as producers of useful information for the state. For a noteworthy exception, Primack and von Hippel 1974. Primack and von Hippel look at the role of science advisors in the United States. They argue that the loss of the science advisor post during the Nixon administration was a loss in part because it forced the administration to learn more about science through funding agencies and the information presented in research reports. They were forced to look at science information more than at science advice. The papers in Elias, Martins, and Whitley (1982) generally present the most interesting sociological discussion of the power and powerlessness of science. The authors of the essays have provided (using a variety of studies of science) an enormous array of evidence on the authority of science and the powerlessness of scientists vis-à-vis the state. See particularly Weingart's essay, ibid., pp. 71–87. Weingart argues quite convincingly that scientists have been used by the state for their expertise and have lost power as a result. He evokes the idea expressed by Elias that scientists have gained power by controlling the "means of orientation" or the intellectual hegemony of scientists. See Elias's essay, ibid., pp. 3–69. While Weingart seems to me to have understood precisely the ambiguity and vulnerability of science's position vis-à-vis the state, he makes one important mistake: he sees scientists as producers of "systematic" *information* for the state. He does not distinguish between developing information and giving advice. Hay also picks up the same theme (with both its virtues and vices) in ibid., pp. 111–19.

There is another view of scientists as "experts" rather than authoritative voices of knowledge that has developed in the sociology of science among the internalists who are relativists, and do not accept scientific truth-claims. See, for example, Collins 1985, especially chapter 1. For mention of the role of scientists as consultants, also see Barnes 1985, particularly p. 48.

4. Recently, Congress has passed some bills giving research money directly to scientific institutions in states not receiving much federal research money. Proponents of these bills have used them to support groups of researchers not favored by the current funding system, thereby skirting the reigning system of control over research. These bills have been surprisingly successful. Their advocates have played on a traditional political contradiction in U.S. science funding: tension between democratic representation and the elitism of science. There has been a tendency in Congress since the nineteenth century to allocate research funds more equally across the states, thereby pleasing constituencies in the name of egalitarianism. In the same way, contemporary politicians, wishing to bring

more resources to their areas, have supported cries to spread government funds for science more evenly across research institutions. See Colin Norman and Eliot Marshall, "Over a (Pork) Barrel: The Senate Rejects Peer Review," *Science* 234 (1986): 145–46.

5. For a discussion of scientific expertise and the state, see Habermas 1970 and the discussion of Habermas in Gaston 1978. See also Bell 1974. For other thoughts about expertise that do not take the political significance of it seriously, see Collins 1985 and Barnes 1985.

The system of communication central to this relationship and hinted at in the text will be discussed in depth in the last chapter, once the character of the relationship between science and the state has been described in more detail. For now, it should suffice to say that the communication of science that "counts" for scientists is text based, as many of the new sociologists of science have pointed out. See, for example, Gilbert and Mulkay 1984; Latour and Woolgar 1979; Knorr-Cetina 1981. Scientific thinking also involves much of the kind of the manipulation of texts that Jack Goody identifies with lists; see Goody 1977. But as Weingart, Primack and von Hippel, and Habermas point out, policy advice from experts is conveyed primarily in a world of oral discourse, where ideas are formulated in a more contextualized way, and they are not made fully accountable.

6. As we will see later, respected research is often not without criticism. It may be countered by other research that reaches other policy conclusions. The important thing is that the science upon which policies are based is authoritative enough so that its legitimacy cannot be entirely undermined by scientific criticism. When it is basically unopposed by other scientists, all the better!

7. One should also notice that NSF proposals routinely point out potential practical benefits of research, as scientists can think them up. This practice is to some extent conventional and hypocritical. For example, scientists routinely cite the potential economic benefits of mining manganese nodules on the seafloor as a reason to fund deep ocean research of one sort or another. Most of these scientists have no real interest in mining these nodules. In fact, some say quite explicitly that they are quite glad that the technology for deep sea mining is so expensive that businesses will not soon be messing around in their research sites. They only mention mining in recognition of the fact that the government expects some returns from its investment in science. They know that mining interests legitimate appropriations for scientific research, so they continue to act in this small way as consultants to the state. They remain (not self-consciously but functionally) a reserve labor force for the government.

8. See Piven and Cloward 1971. It may seem inappropriate to compare people on welfare with scientists, since the latter are so highly regarded and paid, and because they are paid to do work. But one can think of scientists as engaged in a form of upscale "workfare" (workfare being the relief system currently in force for welfare recipients in California). Their high regard is essential to the support for scientists by government, just as the poverty of welfare recipients is essential. The two groups represent extremes on the same continuum. Both are potentially dangerous because of it. As Piven and Cloward suggest, the welfare poor are potentially dangerous because they have few ties to the social order, few com-

mitments to its continuation. Scientists may or may not have this commitment; certainly scientific elites do not need a particular government to do their work. Almost any government would welcome and support them. Thus they have a kind of freedom that is not so unlike that of welfare recipients. The difference is that the potential rootlessness of scientists comes from their importance to countries throughout the world. Atomic physicists who want to defect from the Soviet Union would have no problem finding a welcome in the West and vice versa. One should remember, however, that the bulk of scientists do not have this power, and hence are less valued by governments.

9. Bush 1960. For a discussion of the confluence of industrial growth and military strength, see McNeill 1982, chapters 7–10. For more concrete discussions of the relationship of science to economic and military growth, see Dupree 1986, chapters 7–9, 12–19. See also Russell 1983, chapter 13.

Some scientists I talked to argue that, while labor-force issues dominated early discussions of science policy, they diminished in political salience over time. Increasingly, science was funded because funding had led to such dramatic advances in research. But the science-policy people I interviewed contradicted them. They claimed that it was difficult to justify science funding simply on the basis of its importance to science. In fact, one was quite alarmed by this kind of talk in the scientific community, since it taught younger scientists to forget entirely the political basis for science support in government. Another felt that treating money for science as a form of government patronage was dangerous because it led people to treat science like other forms of political patronage, which it was not. This encouraged congressional allocations of monies to research institutions on purely political bases (pork barreling, see Norman and Marshall 1986). This turn, he argued was bad for science and bad for the United States.

There is also some evidence that manpower issues have remained powerful in getting money from the government for science. The response to Sputnik is a good example. There was a growth in research money for science that was paralleled by growth in monies for science education designed to keep the United States stocked with trained scientists. Manpower issues remain important to the present day. *Science and Engineering Indicators—1987*, put out by the National Science Board (U.S. Government Printing Office, 1987), starts with sections on manpower, goes on to compare the research and development capacities in the United States to those in other countries, and then ends with a discussion of markets for U.S. science and technology. The whole piece is geared toward showing the practical advantages of scientific development, and manpower management is a central part of that strategy. The problem is that this document, as well as others, assumes that education is the only means of manpower development, while this book indicates that it is not.

10. York 1987; Kevles 1978.

11. Piven and Cloward 1971.

12. Anxiety at the end of the war about the happiness of scientists in their home country was magnified by the fact that American science during the war had been so enriched by the contributions of emigrés. The power of scientists also comes from their peculiar relationship to what they study. It is the continuous attention

to making and improving images of nature (both more empirical, descriptive images and more theoretical models) that makes the institution of science socially distinct and powerful. Other groups study the natural world, but usually in quite different ways. Fishermen or generals, for example, tend to have quite limited agendas in mind when they survey the natural world, and many limit the surveying altogether to periods of war or fishing seasons. Scientists, in contrast, make the study of nature an ongoing and culturally organizing obsession. This means that scientists are valued not just because they have information that others need; they have *ways* of thinking about nature that have been acquired purely because they are obsessed by it. Their obsession is socially valued in part because people are often fascinated with nature because it is such a vital part of the human environment and, more pragmatically, because the natural world is the site of the material resources desired by others. The special role of scientists in defining as well as locating natural resources, in developing new means of access to these resources, in defining their relationships to other aspects of nature, and in alerting outsiders to the possibilities of other resources left as yet undiscovered in the natural world gives scientists the power collectively to increase the autonomy of science. And this is true in spite of the fundamental dependence of scientists on government funds and the numerous ties between science and industry. Images or models of nature are a form of cultural capital that help to make scientists social elites and give them the wherewithal to negotiate for power as well as esteem in the social system. For the idea of cultural capital, see Bourdieu 1984; Bourdieu and Passeron 1977.

13. Bush 1960, parts 3 and 6.

14. See, for example, Dickson 1984. William Leiss shows how the pursuit of information is deeply intertwined with the development of research skills in his discussion of the technological basis of modern science. He indicates how the process of doing research is in Western science an *engineering* process, one of using machines to manipulate nature. Thus, both information and the engineering skills used to produce the information result from research. See Leiss 1972, part 2. For another reading of this tradition in critical theory, see Habermas (1981, pp. 366–99) on the critique of instrumental reason.

15. Compare to Piven and Cloward 1971, chapters 1 and 2. I had doubts about the vulnerability of scientists when I began this research. It seemed to me that their social importance gave them so much power that their structural dependence did not make them terribly vulnerable. But the sudden disappearance of DOE funding during the course of my research reminded me of the vulnerability of even those with well-established reputations. The vulnerability of scientists in general to changes in political climate may not be as devastating as dependence on government can be to other groups. Still, some scientists may prosper while others become quite desperate as political events change the funding picture. Focusing on the esteem and security experienced by the most successful scientists draws our attention away from the larger group who experience more of the structural vulnerability of soft-money science. For a discussion of the "Matthew Effect" that helps to protect more successful scientists, see Merton 1973, pp. 439–59.

16. See the history of the National Academy of Sciences by Boffey 1975, where both the consulting role of the organization and its integration into the reward system of science are clearly described.

17. Michael Mulkay describes this phenomenon in his book on the growth of radio astronomy after World War II. He quite rightly points out the connection between wartime uses of science and the growth of new branches of science based on military-developed technologies. See Mulkay 1972.

18. National Science Foundation, *Budget Summary Fiscal Year 1989*, pp. 19, 22.

19. Vetter 1970, p. 16, table 5a 68; National Science Foundation, *Federal Funds for Research and Development*, fiscal years 1981, 1982, 1983, vol. 3; *Surveys of Science Resources Series*, NSF 82-326, table C-84; Schlee 1973.

20. See National Science Board 1987, p. 254, Appendix table 4-20, "Research and Development Expenditures at Universities and Colleges by Field and Source of Funds: 1986," and ibid., p. 286, Appendix table 5-3, "Scientists and Engineers in Academic Research and Development and Basic Research: 1986."

21. Schlee 1973, chapter 9; National Science Foundation 1980; Rona et al. 1984; Corliss et al. 1979; Grassle 1982; 1985; Spiess 1980; Hayman and Macdonald 1985; Jannasch and Mottl 1985; Edmund et al. 1982; Kulm et al. 1986; Johnson et al. 1986.

22. This researcher seems to think that ocean science has an unusual relationship to NSF, but he exaggerates when he claims that NSF functions as an employment agency for oceanographers and not for others. A number of the scientists who lost their DOE funding in the cutbacks described at the beginning of the chapter applied to NSF for research money, and did not get their projects approved. Some are now out of work. So clearly, oceanographers are not given the kind of special consideration at NSF that this man implies. They are as vulnerable as anyone else.

23. Hayman and Macdonald 1985; Jannasch and Mottl 1985; Edmund et al. 1982; Hessler and Jumars 1977.

CHAPTER TWO
THE DEVELOPMENT OF STATE INTEREST IN SCIENCE
IN THE NINETEENTH CENTURY

1. Schlee 1973, pp. 55–58; Daniels 1971, chapters 8 and 12.

2. Daniels 1971, pp. 180–83, 186–92, and chapters 10–11; Deacon 1971, chapters 11 and 14; Schlee 1973, chapters 1–3. Pasteur was also an important figure in the period. For his unusual role, see Latour 1983; see also Latour 1984; Latour 1986.

3. Schlee 1973, pp. 55–58; Daniels 1971, chapter 8; Field 1893, chapter 3; Headrick 1981, part 3, particularly chapter 11.

4. Schlee 1973, pp. 40–58; Daniels 1971, chapter 8.

5. Bright 1898, pp. 2–3, and plates 1, 2.

6. McConnell 1982, p. 15.

7. Lynch 1985a, 1985b. See also Cleveland and McGill 1985, and Mukerji 1985.

8. Bright 1898, pp. 2–5.

9. British Naval Survey 1843. This was also noted in Headrick 1981, chapter 11.

10. Bright 1898, pp. 2, 4; Deacon 1971, p. 298; Schlee 1973, pp. 55–58; Field 1893, pp. 118–21 and chapter 5; Carter 1968, p. 97.

11. Clarke 1959; Schlee 1973, pp. 36–38, 55–58; Daniels 1971, pp. 266–68; Beach 1972, note 55; Reingold 1979, pp. 145–52; Bright 1898, p. 4; Field 1893, pp. 18–21; Carter 1968, pp. 97–98.

12. Schlee 1973, pp. 55–58; Clarke 1959, pp. 31–32; Field 1893, pp. 17–18; Carter 1968, pp. 97–99.

13. Field 1893, p. 19; Carter 1968, pp. 97–98.

14. Field 1893, chapter 5; Clarke 1959, chapter 4.

15. Field 1893, pp. 231–34; Headrick 1981, chapter 11.

16. Field 1893, chapters 8, 9, 12, 15–17; Carter 1968, pp. 106, 134–45.

17. Clarke 1959, chapter 8; Field 1893, chapters 14–16.

18. Schlee 1973, pp. 87–90; Deacon 1971, pp. 306–14.

19. Enclosure in letter from Joseph Matkin to Mother Sarah Craxton Matkin, at sea St. Thomas, on HMS *Challenger*, March 3, 1873, "Substance of Professor Wyville Thompson's Lecture, to the ship's company of H. M. Ship *Challenger*, on the Geography of the sea, & the object of the Challenger Expedition. With remarks on the Progress hitherto made."

20. Deacon 1971, pp. 333–37; Schlee 1973, pp. 107–11. It is interesting to note that these scientists were supported by government partially to make themselves useful as advisors on future cable laying, partially to do some applied work for the Admiralty charts, and partially to pursue their own research program. The tendency to mix all three was apparent even in this early period. Learning about the deep ocean as an exercise in basic research was just not seen as incompatible with providing information and advice. And at least until the cost of analyzing all the data turned out to be greater than the government anticipated, the rationale for supporting science as a way to empower the state was easy enough to sustain.

21. Deacon 1971, p. 350.

22. Deacon 1971, pp. 350–51.

23. Deacon 1971, pp. 388–92; Schlee 1973, chapter 6.

24. Like Chinese Mandarins, scientists in the modern West have developed an esoteric knowledge system. They have used this knowledge system as a way to control access to their own inner circle. They have established a kind of intellectual autonomy that has been the basis of their power. By controlling this information and hence those who could use the information, scientists, like the traditional Mandarins, have elevated their condition by limiting their numbers, and specifying esoteric intellectual powers. In the case of the Mandarins, their intellectual power derived primarily from their training in moral reasoning. Mandarins were thought to be better decision makers because they were steeped in the moral traditions of the society. The power of modern scientists is much more closely tied to their materialist rationality. They may not achieve direct mastery of the natural forces they study, but scientists are masters of skills used in studying nature, scientific method, and reasoning, which give them a culturally as-

sumed superiority in approaching nature for power and profit. See Needham 1969, chapters 4 and 5; Ho 1962, chapters 1 and 2. For the idea of modern Mandarins in the United States, see Chomsky 1969.

Of course, there are clear differences between the two groups. Scientists in the West have often moved outside of government bureaucracy as part of their increasing autonomy. Scientists have had more associations with businessmen and have often developed businesses of their own. Similarly, they have been concerned with things of the hand as well as those of the heart, seeing moral decision as hinging on technical matters, addressed with massive instruments and embedded with many practical meanings. And finally, educational systems in the West have not been dominated entirely by the world view of science, although there has been a clear trend in that direction. At least the scientific elite has not used the educational system as its sole province; thus other forms of intellectual capital is both cultivated and used. See Needham 1969, chapter 5; Ho 1962, chapters 1, 2, 5; Rose and Rose 1969, chapter 2.

What is important to note with this comparison is the way in which both traditional Mandarins and scientists have used knowledge as power. The result in both cases has been careful attention to communication and control of information. Having priority through print is essential to the careers and reputations of individual scientists. Publication is at the heart of the scientific enterprise, just as the ability to write elegant prose was at the heart of the Mandarin educational-bureaucratic system. Not only have communication skills been used for members of both groups to gain prestige within their own circles, but also both groups have protected their information (and hence interests) from outsiders by making the language of their internal communication different enough from the language of everyday life so that the uninitiated are deterred from competing with them or calling their expertise into account. Basic research may be valuable to science (as scientists like to say) because one needs basic principles or theories to make sense of new problems as they arise. And one can also argue (as government bureaucrats funding science like to) that basic research is a source of practical ideas that then are beneficial to the society. Neither of those claims can really be tested. But what is clear is that scientists could have no separate power in society if they were employed by business and government only to solve a succession of practical problems. Chinese bureaucrats may have worked for the government, but they also cultivated their own esoteric knowledge system. Esoteric scientific thought, like classical Chinese philosophy, allows scientists to make special claims about their authority as a group and their character as individuals.

25. Bruno Latour and Steven Woolgar have described in nice detail the function of scientific laboratories as "centers of calculation"; see Latour and Woolgar 1979; Latour 1983; Latour 1986. The importance of laboratories to American science in the nineteenth century is noted by Dupree 1972, pp. 35–36.

26. Latour and Woolgar 1979.

27. Latour and Woolgar 1979; Latour 1986; Latour 1983.

28. Deacon 1971, p. 369.

29. Deacon 1971, chapter 15.

30. Daniels 1971, chapter 12.

31. Schlee 1973, chapter 1; Daniels 1971, pp. 186–92; Rosenberg 1972.

32. Schlee 1973, chapter 1; Daniels 1971, pp. 191–92.

33. Schlee 1973, pp. 27–36; Daniels 1971, pp. 186–89; Reingold 1979, pp. 108–26.

34. Schlee 1973, chapter 1; Miller 1972, pp. 95–112; Beach in Daniels 1972; Daniels 1971, chapter 12; Reingold 1979, pp. 152–61.

35. Schlee 1973, pp. 55–57; Reingold 1979, pp. 145–52.

36. Schlee 1973, chapter 1; Daniels 1971, pp. 191–92 and chapter 12; Reingold 1979, pp. 145–61.

37. Schlee 1973, chapter 7; Daniels 1971, chapter 12; O'Rand 1986, pp. 183–202.

38. O'Rand 1986; Paight 1987; Kevles 1978.

39. Schlee 1973, pp. 206–8.

40. Schlee 1973, pp. 212–13, 217–20.

41. Schlee 1973, pp. 220–22.

42. Schlee 1973, pp. 222–28.

43. Schlee 1973, pp. 206–7.

44. Daniels 1971, chapter 12.

CHAPTER THREE
WAR AND STATE FUNDING IN THE TWENTIETH CENTURY

1. McNeill 1982, chapters 7–9. For nineteenth-century origins of this pattern, see also Headrick 1981, particularly part 2.

2. Rose and Rose 1969, chapter 3; McNeill 1982 chapters 7, 8, 9, esp, pp. 317–45; Schlee 1973, pp. 74–79, 245.

3. Schlee 1973, pp. 246–50.

4. Schlee 1973, pp. 246–49.

5. Schlee 1973, pp. 250–52.

6. Baxter 1968, chapters 9, 10, 11.

7. Baxter 1968, p. 7.

8. Baxter 1968, chapter 1; Rose and Rose 1969, pp. 51–57.

9. See the files of Russell Raitt at the UCDWR, SIO archives, 82–59, boxes 1, 2.

10. Baxter 1968, pp. 174–83.

11. Schlee 1973, pp. 289–92.

12. Schlee 1973, pp. 289–95; Baxter 1968, pp. 51, 180, 184.

13. Schlee 1973, pp. 297–302; Raitt and Mouton 1967, pp. 138–39.

14. Schlee 1973, pp. 304–7; Raitt and Mouton 1967, chapter 13.

15. Baxter 1968, pp. 185–86. The funding history for Division 6 testifies to the importance of the research to the military. Its funds at their height were only surpassed by those given to the missile and radar divisions. See Stewart 1948, pp. 84–97.

16. For an example of scientists' attempts to gain political power, see the history of the Scientists' Movement at the end of the war. Scientists tried to gain some policy power in deciding the application of nuclear power, but they lost the fight. This was pointed out to me by Daniel Paight in his unpublished paper

(1987). See also Smith 1965. The role of scientist as both basic researcher and servant of the state is described in the biographical essay at the beginning of Day 1985.

17. For examples of the reestablishment of control, see Paight 1987 and Smith 1965 on the scientists' movement.

18. Schlee 1973, pp. 134, 143–51. For general information about the continuation of laboratories in this period, see Hewlett and Anderson 1962, pp. 624–35.

19. Schlee 1973, pp. 313–14. For background on Cold War anxiety and scientific research, see Hewlett and Anderson 1962, chapters 16 and 17, particularly pp. 580–82; Hewlett and Duncan 1969, chapters 5 and 6; Sherry 1977, chapter 5.

20. Interview with Roger Revelle conducted for the Becheley Oral History Project, and provided to me by Revelle.

21. Walter Munk papers, SIO archives, 82–57, box 2.

22. Walter Munk papers, SIO archives, 82–57, box 2.

23. Walter Munk papers, SIO archives, 82–57, box 2, folder entitled, "Bikini 1946, General Oceanographic Charts and Reports."

24. Walter Munk papers, SIO archives, 82–57, box 2, folder entitled, "Bikini 1946, General Oceanographic Charts and Reports."

25. Russell Raitt papers, SIO archives, 82–59, boxes 1, 2; Roger Revelle papers, SIO archives, MC6, box 1. The publication rate for the period is gathered from a bibliography compiled by Revelle in 1955. He published two articles in 1940 and 1941 before he was called up for active duty, one on seawater in ground water. In 1943, when he was first on active duty, two more of the papers listed above came out. In 1944 he published the paper on marine bottom samples, and then he published nothing until 1948. In that year he co-authored a paper on Harald Sverdrup (that he dropped from later bibliographies, since it was not a research article). And in 1949 he began to publish on the Bikini research. For the next two years he wrote two articles and then he wrote three and then four in successive years. The description of his military duties is derived from an "Officer Job Analysis Sheet" filled out on July 7, 1943. There he is described as devoting most of his time to the deployment of the Bathythermograph (MC6, box 1, no. 28), and from the myriad travel duty sheets from 1943 to 1945 (MC6, box 1, no. 30). A 1963 bibliography of Revelle also lists three declassified papers written during 1942 and 1943. They are on sound ray patterns in seawater, the study of surface ship wakes (apparently for acoustical identification), and radar wave propagation. In his unpublished work, Revelle writes useful reports for the Navy on acoustics; but these pieces could not be published, so exploitation of applied research to serve the interests of science could not really be accomplished in this period.

26. Russell Raitt papers, SIO archives, 82-59, boxes 1, 2. Compare to Munk and Revelle papers mentioned above. See also Paight 1987 and Smith 1965 on the American Scientists' Movement. See also Rose and Rose 1969, chapter 5, on British scientists in this period.

27. See Deborah Day's description of Revelle at the beginning of her *Guide* (1985).

28. For the Bikini test, see Schlee 1973, pp. 314–15; Hewlett and Anderson 1962, pp. 580–82.

29. Before the ONR was started, an effort was made to develop a research board in government, sponsored by the military but not dominated by them, that would oversee scientific research. There were some disagreements among military leaders about the wisdom of this, but it seemed unlikely that scientists would remain in the military to continue research when there were opportunities in universities and private research corporations. So some means had to be found to involve scientists in research while they had other outside obligations. The research board failed as a political entity, so it did not develop a solution to this problem. This left the problem to the military itself, and it was handled by the Navy through ONR. See Sherry 1977, chapter 5; Mangone 1977, p. 17.

30. Sherry 1977, chapter 5; Hewlett and Duncan 1969, pp. 58–68.

31 Sherry 1977, chapter 5; Mangone 1977, pp. 51–64.

32. Sherry 1977, chapter 5; Mangone 1977, pp. 46–60.

33. Bush 1960. For a discussion of the confluence of industrial growth and military strength, see McNeill 1982, chapters 7, 8, 9, 10.

34. Bush [1945] 1960; Schlee 1973, pp. 311–12.

35. Bush [1945] 1960, p. vii.

36. Bush [1945] 1960, pp. 18–19.

37. Angela O'Rand documents this process. European scientists were involved in this kind of "purification" of science earlier, certainly by the eighteenth century. The misleading character of this view of science is documented by Bruno Latour. Latour (1986) argues that scientific research in the present, work that is in progress, has one language, a language of controversies and polemics, and that canonized science has another language, one that idealizes scientific knowledge, treats it as truth, and implies that new scientific research simply adds to knowledge in a cumulative fashion. It is this second language, taken from the elite vision of scientists, that is used to justify state research funds.

38. Bush [1945] 1960, p. 34.

39. The image of a reserve labor force, occasionally mobilized to benefit the state, is too threatening to the elite status of scientists to be stated explicitly, so it is never quite on the surface of this paragraph but neither is it far from it. Perhaps in the world of the military, state interests are clearer and thus more likely to poke through rhetorical stances. But whatever the reason, we see hints here that scientists are experts who can be useful to the state as consultants.

40. Bush [1945] 1960, p. 33. There were also some unwritten assumptions in this system. One was that scientists would apply for funds competitively. A corollary of that was that the best of them would be favored by the system. They would not always get the most money (for they might not need the most), but they should get the best access to funds for their work. Moreover, those scientists pursuing basic research would be favored over those doing applied work (at least by this agency or its progeny).

41. The discussion of the manpower needs of the state is usually not explicit in the Bush text, but it is expressed openly in the section of the Bush document entitled "Renewal of Scientific Talent." This part of the proposal argues that there

is a deficit of trained scientists resulting from the war (due to both a lack of protection of scientists from the draft and the disruption of education during the war). James Conant, then President of Harvard University, is quoted as saying:

> . . . in every section of the entire area where the word science might properly be applied, the limiting factor is a human one. We shall have rapid or slow advance in this direction or in that depending on the number of really first-class men who are engaged in the work in question. . . . So in the last analysis, the future of science in this country will be determined by our basic educational policy. [See Bush 1945, 1960, p. 23]

There is no discussion here of how the funding system as a whole will serve manpower needs, but a section on fellowships and scholarships says that student support is needed to encourage students to enter science and medicine. Also, in Appendix 3 the Committee on Science and the Public Welfare goes further, tying student support to faculty research. It argues that basic research in the universities is important both for improving the level of scientific research in the country and for providing a first-class education for students. This committee also recommends fellowships for mature scientists who have been doing applied research. They are given the money to spend time at the universities doing basic research to facilitate application of basic to applied work. See Bush [1945] 1960, pp. 70–134.

All these recommendations focus on manpower issues of one sort or another. The major rationale for this is cultivation of talent, but one can also see it as a means for improving the research skills of scientists deemed desirable by the state. In this way, the Bush report becomes explicitly devoted to manpower development, even though it is too tactful (or politic?) to discuss the work of distinguished scientists in these terms.

42. A scientist who specializes in acoustics put it this way:

> There was a program to develop a missile with a submarine that could sit underwater, shoot the missile out of its torpedo tube. It would go up into the [air], light off and fly as a rocket like that for some distance—twenty or thirty miles, something like that—go back down and then explode, hopefully in the vicinity of some other submarine. And the only way you know where the other submarine was, was by using acoustics. And so they were concerned about how accurately you could tell, if you decided what direction your target was, how accurately you could really know that, given the environmental concerns and the nature of water and the nature of the seafloor, if the sound kind of bounces. And so of course this was a pretty big program. There were these fairly fundamental acoustics questions, and so they pulled together an advisory committee. And in this case there were eight of us. One each from several of the Navy labs that were involved in this sort of thing. One from [Blue Lab], one from [Right Lab]. [Right Lab] had a laboratory that was sort of like [ours] except they were more involved with torpedos than work with the physical environment.
>
> . . . The laboratories at [Wet Lab] and [Right Lab] were primarily funded by the Bureau of Naval Weapons, which was concerned with torpedos and that sort of thing. And they were involved. They were both laboratories that had been established during World War II and had continued on after it. There was a pretty good underwater acoustics

community sort of network. People who knew how to make weapons were acoustics people, and knew the acoustics people in other areas as well.

The relationship between "pure" science and its applications for defense is also discussed by Sherry 1977, chapter 5, where he quotes James R. Newman speaking about scientists' reluctance to serve the military directly.

There is a strong element of morality and personal ethics which appears to the average scientist. He will work quite readily on a synthetic product of an electro-magnetic device [sic], or on a new type of atom disintegrator [sic] even if, subsequently, the results of his research . . . turn out as useful military weapons. But, if he is told to work in an ordnance laboratory or arsenal, he will feel that he is devoting his entire life to the making of lethal weapons and, except in war, he will reject the suggestion.

Sherry comments on the passage: "Many scientists did not possess sensibilities which required such subterfuge, and others would not have been fooled by it. But it made sense to make research as intellectually challenging to scientists as possible."

43. Schlee 1973, p. 314. See also Hewlett and Duncan 1969, chapter 8, for information about increased AEC funds for research and new funds for basic research in the period. But there is no specific mention of ocean research in the official histories of the AEC.

44. See the John Isaacs papers on Operation Crossroads and later AEC projects, SIO archives, 81-96, boxes 2–6, and Walter Munk's papers on Crossroads, SIO archives, 82-58, box 2.

45. Wenk 1972, pp. 41–44. For a general analysis of science policy in relation to basic research, including discussion of the Sputnik crisis, see Dickson 1984, particularly pp. 27–28.

46. Spiess 1987, particularly table 1 and figure 1.

47. There has been a tradition in NSF of trying to support research with some potential economic usefulness. Scientists have followed this charge to the agency, and have done work that assesses the economic value of exploiting certain natural resources. Often this work does not lead to direct economic benefits. See, for example, Manheim 1986, and Broadus 1987. These studies find that the commercial potentials of seabed mining are still not great enough to warrant serious business pursuit.

This kind of research is prevalent among marine geologists, some of whom are seriously interested in economic exploitation of seabed resources, but many of whom say this in order to get funding for basic research. One man who had talked about the commercial potential of the hydrothermal vents even went so far as to say that he did not want any corporations messing around in his research site. He and others like him simply talk about the value of their research in these terms in order to help secure research funds.

48. Wenk 1972, p. 341; Dickson 1984, pp. 122–24.

49. See Cole, Rubin, and Cole 1978. These authors claim that peer review in NSF is highly bureaucratized and based on the merits of individual proposals. There are also some indications of discomfort with the proposal orientation. The critics of peer review are both numerous and come to their criticisms from quite

different perspectives. See, for example, Shaply and Roy 1985, who do not believe in the social value of basic research. See Stewart and Feder 1987 on the failures of peer review (primarily in journals but by extension elsewhere) to distinguish between good and bad research (including fraudulent research). And see also a plea for more funding of individuals rather than projects in an editorial in *Science* by Daniel Koshland, Jr. (1985).

These criticisms of the peer review system are balanced by outrage at the lack of peer review in the distribution of funds in government labs. This issue was introduced by Frank Press in a "Perspective" in *Science* 231 (1986): 1351, and stimulated a large outpouring of letters. See *Science* 232 (1986): 563, 919, 1183.

Science has also recently published expressions of outrage because Congress assigned money to scientific establishments on a political basis, without any concern for the quality of science done there. See Norman and Marshall 1986.

50. Cole, Rubin, and Cole 1978, chapters 3 and 5.

51. Cole, Rubin, and Cole (1978) argue against the idea that some scientists gain cumulative advantage over time in this system. But they do this by distorting the concept of cumulative advantage. While Zuckerman argues that well-known scientists use their visibility as well as ability to increase their intellectual skills and research resources, Cole, Rubin, and Cole think that scientists are advantaged only when they can get funding without writing better proposals. But the point is that if they have a better education, more research opportunities, a job at an institution with more resources, access to better graduate students and post-docs, access to better libraries, and so forth, they are better situated for writing good proposals. See Cole, Rubin, and Cole, chapter 5.

CHAPTER FOUR
MANAGING THE SCIENTIFIC LABOR FORCE

1. On the decentralized nature of research and science advice, see Golden 1988.

2. Using NSF and other agencies as a source of information about who is doing what in the scientific community is not without problems. Primack and von Hippel suggest that one reason why the position of science advisor is so important in the administration is that strategically placed scientists know more about what is going on in science than agency personnel. Primack and von Hippel argue that advice is better ascertained from within the community of scientists in part because of the secretiveness of scientists and their distrust of outsiders. This argument parallels the one made in this book about the greater importance of advice over scientific information for the formation of policy, also stressing the greater value of informal over formal communication.

3. For a discussion of surveillance, see Foucault (1979, part 3, chapter 3) on panopticism. It is important to remember that the information used for practical purposes is embedded in the *expertise* of scientists rather than in a body of information that the government uses itself. Compare to Weingart in Elias, Martins, and Whitley 1982.

4. Bush [1945] 1960.

5. For as description of the growth of NSF and its relationship to other agencies, see Wenk 1972, chapter 2, particularly pp. 46, 57–58.

6. Scientists doing applied work are employed explicitly to advise the state; scientists doing "basic" science are asked for advice only when they have the highest regard as scientists. Thus, ironically, providing consultation can have two polar meanings: at the highest levels it is seen as an honor, while at the lowest levels it is indication of a lack of autonomy and prestige. See Primack and von Hippel 1974, and C. Hay in Elias, Martins, and Whitley 1982.

7. See Dickson 1984, chapter 2, and Nelkin 1984.

8. For general descriptions of big science, see Derek de Solla Price 1986 and Hagstrom 1965, chapters 2 and 3. See also Erzkovitz 1984.

9. The tapes are surprisingly useful for this purpose because of the way they isolate two channels of perception and communication. The visual images show us not the scientists in the submarine, but rather the ocean environment; the soundtrack, on the other hand, only provides information about what is going on *inside* the submarine, what is said and how the machinery sounds from that position. Thus the images seem to present nature, while the audio certainly presents evidence of the scientists' culture. The division is not that clean, and what you see on the tapes is only roughly what the scientists see and certainly nowhere near what natural forms are actually around the sub. But the effect is striking and helps to isolate what the scientists are doing from what they are seeing. Machine noises emitted by the sub, the talk of the scientists about the work, radio discussions between sub and the mother ship, and even in some cases music played by the scientists during the dives are heard while you see vast stretches of seafloor roll by.

Listening to the scientists talking "scientifically" about what they are seeing, one begins to get a sense of how they use both human language and the language of science. And hearing the noise from the submarine itself, one becomes more aware of the submarine as a piece of familiar culture that carries with it laboratory paraphernalia and a definition of the dive situation as a laboratory setting. The noise, lights, experiments, and propellers also serve as a reminder of surface culture, since they are familiar parts of a surface culture brought into the deep. This kind of emotion management has been talked about as "feeling work." See Berger 1981, and Hochschild 1983.

10. For descriptions of Project FAMOUS, see Heirtzler 1974, 1975; Ballard et al. 1975; Bellaiche et al. 1974; Ballard 1975. One might argue that this research has been spurred by an interest in understanding mineral deposits on the seafloor for potential mining, but that is pushing it a bit. The whole idea is not yet taken too seriously by researchers because it is clearly not cost effective to mine in the deep ocean. One could also argue that the Navy has interests in the ridge crests around the world, and these surveys contribute to their strategic knowledge, but that would be pushing it too. Of course, they are interested in this information to some extent, but that does not account for the huge budgets allocated for research with *Alvin*. So what does explain it?

11. Letter from Nuclear Research Company (pseudonym) to the Atomic Energy Commission, August 1956.

12. Appendix 1, p. 30, proposal, 1956.

13. Like other relationships, the funding one is based on the hope or ideal of

mutual trust; both parties must seem sincere and reliable. What this means is that scientists must describe themselves indirectly with their prose as reliable allies of their funding sources. They may be crazy as loons and do all their work in their pajamas when all the other members of their laboratory are asleep. But that does not matter. This kind of behavior is not put in public documents that would affect a sponsor; moreover, it indicates an innocence about worldly affairs that can make a scientist seem all the more reliable to a worldly partner. What is more important for research proposals and reports is that scientists perpetuate the ideal of scientific dispassion. Scientific researchers are equipped by training to maintain a nonthreatening stance; they are taught to try to be objective and not become personally invested in any particular interpretation of the research process and findings. In this way, they almost automatically cleanse their subject matter of political or moral issues that might make funding agents nervous. When discussing politically charged topics, their prose (in papers as well as proposals) emerges with a massaging quality, turning everything into a technical or empirical issue. This makes their writing seem devoid of passions other than intellectual ambition. The mythological scientist—pure mind in pursuit of pure theory—is the major character even in research proposals for applied projects. An early member of ONR said that he and his peers avoided funding projects that were trying explicitly to please the Navy. Presumably, a fawning partner is seen as unreliable and not someone who will maintain the proper role for science. For discussion of the gender relations implicit in this kind of dispassion, see Keller 1984.

14. For a discussion of voice, see Bakhtin 1981. See also a variety of structuralists and post-structuralists on the social nature of talk: Harari 1979, particularly Foucault, pp. 141–60; Kurzweil 1980; DeGeorge and DeGeorge 1972; Eagleton 1979.

15. The Principal Investigator who eventually developed the camera system was among the men who had written the original scientific proposal to the AEC in 1956. He had been one of the AEC's bright young men at that time because he had been such a visibly gifted problem solver in the Bikini tests, and was being watched to see what more he could do. He continued to work with the AEC on atomic tests and became a regular recipient of their funds, as well as ONR block funds. He made such a mark as a technological innovator that he was a major figure in ocean engineering for both applied and research purposes, and he was one of the few engineers deemed capable of doing "real science." So the AEC had every reason to trust him as a helpmate and as an inventor of research instruments.

He seems to have taken advantage of his good reputation to initiate the underwater camera project—a project he had long wanted to try. He realized that the AEC wanted inexpensive and efficient means for monitoring life and physical processes on the seafloor, and he seems successfully to have convinced them that his interests coincided with theirs. The result was the development of a series of cameras to go on a free vehicle that he had already invented for making temperature, current, and other readings down on the seafloor.

16. This perhaps explains the relative equity of NSF found by Cole, Rubin, and Cole (1978) in the evaluation of proposals. Compare to the thesis proposed by Jackson about bureaucracy and gender. Everyone has the same chance but not

the same skills. What you don't see are cumulative advantages. See, for examples of how this inequity affects women, Gornick 1983.

17. See Nelkin 1984 chapter 5; Primack and von Hippel 1974.

18. There is a lot of research in the sociology of science now on the creation of alliances among researchers as central to the success of the scientific ideas. It is a little difficult to say precisely what research serves science and what serves the state more, but one can identify moments when allies leave a research project because they see it as weakening their positions. See Latour and Woolgar 1982, pp. 35–43, and Collins 1982, pp. 44–64. See also Collins 1984, and Latour and Woolgar 1979.

19. There are obvious differences between analyzing this kind of criticism of scientific projects as expressions of a normative system or part of a political one. (For examples of normative analyses of science, see Merton 1973.) The political image of scientific behavior assumes that scientists mobilize the former in order to serve their interests. It does not (as some suggest) deny the existence of a normative system. It just assumes, as Bennett Berger (1981) does, that cultural systems are used and invoked only when circumstances make them salient. The political process is a major way in which this kind of salience is established. For discussions of the political nature of science, see Hagstrom 1965; Nelkin 1984; Latour 1984; Latour 1986; Primack and von Hippel 1974; Elias, Martins, and Whitley 1982; and Dickson 1984.

20. The Bush system of scientific funding was clearly served by this project in spite of its eventual demise. Intellectual freedom was granted for quite a while so the scientists could develop and learn to exercise some skills. It lasted as long as the project could be justified as serving *science*. When the results lost their scientific value, the project was rejected; the definitions of proper scientific activity to be supported by government were sustained.

It is easy enough to recognize how skills of the sort developed during this project might have potential use for its sponsors. After all, the project was to some extent an applied one. The development of the camera was justified on its scientific applications, but its practical applications for the AEC were not far under the surface. It is clear that the AEC would have use for an inexpensive but scientifically respected camera system for monitoring the deep ocean, and it supported the research group as long as it seemed to be producing such a beast. But the usefulness of the system depended on the scientific evaluation of the project that tied its applied purposes to criteria of quality derived from canons of basic research.

CHAPTER FIVE
LIMITS ON THE AUTONOMY OF SOFT-MONEY SCIENTISTS

1. Spiess 1987, pp. 14–15.

2. Cumulative advantage (the "Matthew Effect" in Merton 1973) also has its limits. For one thing, it only affects a minority of the scientists working at any given time. Most scientists work with *cumulative disadvantages* rather than advantages that make them vulnerable to the power of others. They may get funds, but they have to spend large amounts of time writing unsuccessful proposals be-

fore they put together one that works. They will then use the results to write articles that have no particular effect on the field. These people may sometimes be called upon to review proposals because of their specialties, but they will not be part of the small group of scientists routinely called upon by agencies for consultation. Some of these people are simply too young to have had much opportunity to gain advantages. Younger scientists are, after all, faced with finding a position for themselves in a research map already drawn by their senior colleagues and funders. But many more are mature scholars who do adequate, but not noteworthy, research and publication.

The problems of doing soft-money research (perhaps not surprisingly) are most daunting when scientists begin their careers, but they do not disappear as scientists get more established. In fact, one of my younger informants told me that it is the older university scientists with established but not outstanding reputations who often complain most vociferously about funding problems. Established (but not famous) researchers with high-level appointments but no guaranteed salary feel that they should not have to live with that kind of insecurity; they ought to have some of their salary guaranteed after awhile.

Between younger researchers and their cumulatively disadvantaged senior counterparts, there is a clear majority of scientists who are more vulnerable than powerful within the funding system. The autonomy of distinguished scientists may also not be as great as expected. Their proposals get turned down, too, if they do not please reviewers. They sometimes find that they cannot continue an old line of research because it is seen as passé, yet they cannot take a new direction because they are not seen as competent to do a new kind of research. As Cole, Rubin, and Cole (1978) found out in their study of NSF, reviewers are always attentive to the possibility that a distinguished scientist has gone "over the hill." In different ways, then, scientists continually have to prove themselves. So the autonomy that scientists *do* have (even the most successful of them) may be less than conventional wisdom would lead us to expect.

3. Where agency personnel have already developed this helpful attitude, even younger researchers can take advantage, arguing that their work, too, is contributing to some important new directions of their fields.

> It was just purely an accident that [I got ONR funding so easily]. There was an advance in analytical chemistry, a technique called flow injection analysis which is the principle on which this instrument runs, and I had just come across a review article on it, and I knew that ONR was interested in this field.
>
> . . . [At ONR] they're interested in, among other things, just basic science. They don't have the resources, nearly as many resources [in the Navy for research] as for weapons, and they like to make an impact in an area that they feel is being neglected. And one of the things they've identified is analytical, development of new analytical techniques for [chemical] oceanography, and so they're funding at least five groups I know of. [He names them]. In fact, they're being extremely successful at it, and I think in, you know, three or four or five years down the road you'll find that oceanography is using techniques that have been funded by this sort of current cycle of ONR initiatives.

This man is able to get ONR support for his work, in spite of the fact that he is a young researcher because he has identified and can support an agenda set up by the agency. He is appropriately boosterish, knowing how to sell a program of research that serves his interests.

4. The alliances among scientists noted as part of the intellectual combat in science (see, for example, Hagstrom 1965; Latour and Woolgar 1982; Collins 1982; Gilbert and Mulkay 1984; Hay 1982) are paralleled by the alliances formed among researchers to facilitate funding. Are such groups the same? Judging from my interviews, it seems there is a great deal of overlap. Intellectual enemies do not like to share the same ship, but occasionally they are asked to.

5. This situation has the same kind of stability Wallerstein describes in the modern world system. Where political and economic units are the same, the economy is dependent on the success of the political unit. For scientists, having funding from one source makes them dependent on their success with that agency. If they have multiple funders, the loss of one source of funds can no longer cut off the project, just as the destruction of a state in the world system does not disrupt the international economy. See Wallerstein 1976.

6. Spiess 1987.

7. For descriptions of Project FAMOUS, see Heirtzler 1974, 1975, Le Pichon 1975; Ballard et al. 1975; Bellaiche et al. 1974; Ballard 1975.

8. For a general discussion of plate tectonics in oceanography, see Menard 1986; Cromie 1980. For a discussion of its role in the history of oceanography, see Schlee 1973, chapter 9, and Shor 1978, pp. 308–16.

9. Colin and Marshall 1986.

CHAPTER SIX
TECHNOLOGICAL DEPENDENCE OF SCIENTIFIC RESEARCHERS

1. For a description of the research fleet, see "The University-National Oceanographic Laboratory System: The Research Fleet," in Cullin 1983. See also *Oceanus* 25 (1982), a special volume on ships in oceanographic research, especially the introductory essay by Spencer 1982.

2. For information about submersibles, see Vine 1982 and Allmendinger 1982. For a historical description of the equipment used for underwater research, see Heberlein 1971–72.

3. Fred Spiess describes the development of Deep Tow in Sears and Merriman 1980, pp. 226–39. The equipment itself is described in Spiess and Tyce 1973. Some of the work done with Deep Tow is described in Spiess and Mundie 1970, pp. 205–50; Spiess et al. 1969. For descriptions of Angus, see Phillips et al. 1979.

4. The Argo-Jason system as a concept is discussed in Ballard 1982. The articles on the *Titanic* discovery are too numerous to cite. It was described briefly in *Science* in the "News and Comment" section in 229 (1985): 1368.

5. For early description of the RUM series of vehicles, see Anderson and Todd 1960; "Background Information on the Sea Floor Work Vehicle RUM," enclosure to MPL file 02-U-30 (January 25, 1972). For the second phase of RUM, see Alexander 1975a, b.

6. For a description of early free vehicles, see Ewing, Worzel, and Vine 1979, chapter 1. For more recent use of these vehicles, see Isaacs and Schwartzlose

1975; Hessler et al. 1978. For a description of the use of fish traps on free vehicles, see Schulenberger and Hessler 1974.

7. For information on transponder navigation, see, for example, Edgerton and Udinstev 1973; and Milne 1979.

8. See *Oceanus* 24 (1981).

9. Lighting problems are discussed in *Oceanus* 18 (1975), a special volume devoted to undersea photography techniques.

10. Military *techniques* as well as technologies have affected science. There has been a long-standing interest by oceanographers in using deep sea photographs for biological analysis, but there are numerous problems in interpreting the images. This problem has been less for deep ocean geology, where pictures (primarily sonar ones but also photographs) have been used to define bottom features that fit or contradicted the reigning version of plate tectonics. With marine biology, the questions and answers sought from the deep have been more difficult to address with photographs. If you take a picture, how can you generalize from it to estimate population size? How do you sample with a camera? One solution developed (apparently successfully) was to apply techniques of aerial photo analysis, developed by the military, to photographs from the seafloor. Here science again looked to the military to improve scientific research, this time using the military's greater experience in analyzing these kinds of data. See Brady 1982; see his reference to a Mahoney article on Defense Mapping.

11. Industry supplies about half of the research and development money available to science and engineering in the United States. See Office and Management and Budget 1988.

12. Schlee 1973, pp. 358–62; Shor 1978, pp. 308–16.

13. For a description of the technology being developed for use primarily by the oil companies, see "RVC: Expanding Role Offshore," *Northern Offshore* 6 (1977): 70, and Benthos Undersea Systems Technology Product List, October 1983. See also Ballard 1982 and Allmendinger 1982, pp. 24–29.

CHAPTER SEVEN
TECHNIQUES AND STATUS IN SCIENTIFIC LABORATORIES

1. This makes technicians a kind of support personnel in the laboratory much like the support personnel in art described by Becker (1982) in *Artworlds*.

2. Shapin 1988.

3. For an illustration of this attitude and the definition of scientists as the new aristocracy, see Rowland 1979.

4. Bush 1960, pp. 18–19.

5. The view of technology as an extension of human capacities (and as a source of progress and human betterment) is a common one in the culture, and it is in direct opposition to another common one: the view that technology is a means of deskilling the work process, making those who work with the machinery mere mindless extensions of the people who organize their work. Both perspectives on machinery are apparent in the world of ocean scientists.

6. Simmel 1950b.

7. Science to most people seems a very alien world, mostly because of the

obtuse language of scientific debate and the mathematics used by many scientists, which are well beyond the training of most ordinary readers. But there is another alien characteristic of science that people rarely think about. Scientists work with an array of machinery that is as foreign to most observers as the language of science is to most readers. The fundamental skills of scientists that help to distinguish them from the rest of the population are their abilities to penetrate these two cultural walls. They learn how to read the written material, and learn how to use the odd arrays of machines. They combine these specialized skills with organizational and other practical ones to solve the intellectual puzzles and sometimes practical problems that are the focus of their research.

Many distinguished and competent sociologists have analyzed the skills and strategies used by scientists in scientific writing, so we know something about them. But relatively little has been written about the technological and practical skills that scientists combine to make their experiments work. This is one reason this chapter will focus on the latter. The other reason is that technology is so palpably important to the functioning of scientific labs. It is the central form of material culture in the world of science, directing and limiting the activities of scientists in their research. Scientists could not imagine themselves without it; and neither should we. See Latour 1986.

8. Ellul 1964.

9. Sudnow 1978.

10. Compare Ellul to Sudnow.

11. Ellul 1964. For the gender significance of the idea of technique, see Keller 1984.

12. Sudnow 1978; Latour 1985, 1986.

13. Latour 1986 and to some extent 1979 on black boxing.

14. The *Oceanus* (18 [1975]) issue on deep sea photography, for example, illustrates some different signatures developed with photographic equipment before 1975.

15. Revelle interview from the Oral History Project, Berkeley, California. This is taken from an early draft of the transcribed interviews so it does not correspond (necessarily) with the final edited version of the interviews. It was given to me personally by Revelle.

16. Collins 1984; Latour and Woolgar 1984; Hagstrom 1965.

17. The citations made within the text have been omitted because they would give away the identity of the researcher who gave me the proposal. Even though most marine researchers may be able to guess the author, this at least provides protection from more general recognition.

18. See Allmendinger 1982.

19. For discussion of the social position and treatment of artists, see Becker 1982. For problems that result from crediting senior scientists with work they have not done, see Stewart and Feder 1987.

20. Becker 1982.

21. Compare Hampton 1970 to Wollen 1969.

22. For a serious example of this approach, see Hall 1963. This perspective is particularly visible in the history of science, but it also creeps into the sociology

of science as well. See, for example, Zuckerman 1977; Merton 1973. This work is organized around the concept of cumulative advantage, trying to see how some individuals succeed more than others. It does not take greatness as a personal attribute without social dimensions, but it does make the lives of a few individuals still capturing the world of science.

23. See Bush 1960, pp. 18–19.

24. Bush [1945] 1960.

CHAPTER EIGHT
EXPANDING THE DOMAIN OF SCIENCE

1. Natural scientists may specialize in studying the world's creatures, materials, and processes, but they are not alone even in this. Definitions of the natural world are multiple at any given moment in history and change over time. Scientific investigation and commercial exploitation of nature have created the peculiar modern images of nature, but they do not have absolute power. They exist alongside superstitions, religious explanations, and various secular folk traditions. Any longtime fisherman would claim mastery over secrets of nature that scientists do not know. Naturalists routinely contend that scientific understanding of our physical environment is seriously limited. And so do many religious leaders (particularly fundamentalists). Even dominant political or economic institutions like the military and corporations, in spite of their fundamental allegiance to scientific thinking, often find science less help than they hope for solving their practical problems. For an example of the confrontation of different views of the ocean, see Callon 1986, pp. 196–235.

This is to say that there are many voices laying claim to parts of the natural world as their intellectual/spiritual province. And to understand the power of science as an institution, we must understand that there is competition for our imagery about the natural world. See Gusfield 1981 for other examples in the sciences. To the extent that scientists seem the sole definers of the natural, they are simply being successful in making their imagery seem more legitimate than that of competing groups. What is important for studying science, then, is to understand how scientists use the research process to increase the domain of science. See Bourdieu and Passeron 1977 on cultural capital. See also Bourdieu 1975. The extent to which the "capture" of new domains by scientists with their models affects their credibility and what this means for labs is illustrated in Latour and Woolgar 1979. See also Williams and Law 1980.

2. For discussion of the nature/culture line, see Leiss 1972; Thomas 1983. For discussion of collection, see Mukerji 1983. For collecting as a social form in Western homes and gardens, see Thacker 1979; Strong 1979; Bermingham 1986; Brockway 1979; Girouard 1978, 1979.

3. How did scientists come to gain enough power to be interesting to the state and yet able to remain apart from it? Keith Thomas's historical analysis of the beginnings of Western science is a useful place to begin. He suggests that prior to the "scientific revolution," Europeans subscribed to the Christian conception of nature as the domain given Adam in the Garden of Eden. Nature was made antagonistic to human domination after the Fall, but it still remained man's do-

main. The scientific world view changed this. Nature no longer inherently "belonged" to man to be used as he pleased. It could be used to serve human interests only when someone could unlock its laws and make them available for human manipulation. Scientists were the designated group capable of reappropriating nature for human domination, and, with this charge, they made the natural world their intellectual property. Of course, they could not make nature literally their property, but by making models of the natural world, scientists could claim they had uncovered natural laws. And if they could comprehend natural law, they could claim nature as somehow at their command. (See also Leiss 1972.)

In his study of Pasteur, Latour (1984) demonstrates some ways scientists have used "nature," meaning their models of nature, as a source of power to extract resources from the state. Pasteur was able to make his theory of disease a basis for founding a major laboratory because the success of his public experiments showed his control of natural forces. These natural forces became socially "his," and those wishing to make them less threatening to society had to accept his claim to them as his intellectual property.

Using a different vocabulary, we can say that the power of science in modern societies comes from a double alienation, first an alienation of nature from the human sphere (with dualism), and then an alienation of natural forces from the separate domain of nature through its reappropriation in scientific models. Nature is reconstituted as an "ally" for scientists to use as a tool for gathering resources from others (both insiders and outsiders to the world of science). See Latour 1984 and 1987. This double alienation makes the natural world (at least potentially) the intellectual property of scientists. This "control" of nature is what made science originally of interest to states and has served as a source of empowerment in the modern world.

For a discussion of consumerism and the European approach to property, see Mukerji 1983.

4. Baxandall 1985. Barry Barnes also uses the concept, but in a different way, in Barnes 1983.

5. What does it mean to alienate nature in order to empower science? It certainly cannot mean alienation in the social psychological sense, since that requires a kind of consciousness. Nature can obviously not feel alienated. It is not fair to say that a dog, for example, feels alienated when it is the object of research. It is more likely to feel boredom or pain than alienation. But if alienation is a structural relation, where the interests of creatures or objects are made subservient to the interests of researchers, we can certainly say that when Harvey studied the circulatory system by letting the blood drain from live dogs, he was alienating the dogs from their purposes. For a description of Harvey's work, see Whitteridge 1971. When plants are ground up for chemical analysis, you can also say that this is a form of alienation, since it disrupts the normal life course of the plants and makes them into an artifact of science. And when rocks are taken from a scarp and moved into a laboratory, this is a detour from the rock's normal progress and a recontextualization or reappropriation of them in the human world. So again one can legitimately talk about an alienation of nature.

6. Compare to Goffman's concept of framing in Goffman 1974 and Goffman 1976.

7. See Leiss 1972.

8. Latour and Woolgar 1979.

9. See, for example, Latour 1983.

10. See Sternberg 1985.

11. Some of the major discoveries since the Second World War include the exploration of the mid-Atlantic ridge and the study of spreading centers (particularly in the Pacific). See Schlee 1973, chapter 9, and Cromie 1980.

12. See Schlee 1973, particularly chapter 9.

13. See Edmund and von Damm 1983, particularly pp. 78–81; Lupton and Craig 1981; and Macdonald and Luyendyk 1981, especially pp. 100–102.

14. These tapes are available for study in the archives at the Woods Hole Oceanographic Institution.

15. Corliss et al. 1979; Grassle 1982, 1985; Jannasch and Mottl 1985; Childress, Arp, and Fisher 1984; Hessler and Smithey 1984.

16. Edmund and von Damm 1983.

17. The structuring of scientists' observations with signatures and routine measurements is all the more useful for scientists who are trying to make sense of a novel site like the hot springs. Much of the deep ocean biology done prior to the discovery of the vents, for example, focused on organisms living in the silt layers along the vast plains on the seafloor. The vent areas were mountainous, some of the creatures they found there were often unknown to them, and all the animals were huddled around scalding water seeping or soaring out of the rocks, carrying massive amounts of toxic chemicals. If scientists could not explain what they saw they could still use signatory techniques. Ecological biologists spent a great deal of time simply labeling and counting animals, doing a census to see who lived there. They applied routine scientific approaches to biological problems so they could answer basic scientific questions. Other scientists did more specialized experiments on metabolic rates and the processing of toxic chemicals by the vent animals, using signatory instruments and techniques developed from earlier work with "plains creatures." Some researchers found their existing signatures inadequate or inappropriate for this work, and revised them as they learned more about the vents. Through the application of techniques, researchers carried themselves through the three stages for the alienation of nature: framing (simple ostensive modeling), making data (advanced ostensive modeling), and theoretical modeling. In this way, they made the vents into images in the "big picture," and used them to make new and bigger pictures, thereby claiming once again a part of the natural world for science.

18. Turner 1977, pp. 13–19.

19. For individual scientists, trying to learn something new through vent research had potential cognitive, personal, and social risks. Just like the intellectuals in Europe, after the discovery of the New World, who had difficulty absorbing all the novel plants, animals, and cultures there, these scientists have had difficulties seeing and modeling patterns in nature that they could not understand with familiar cognitive categories. For a description of expansion and the intellec-

tual crisis in Europe, see Elliot 1970. All of these scientists had to begin their work by developing an appropriate (whatever that means) set of categories.

These activities by the vent researchers may seem unusual, since scientists rarely face such baffling novelty in their research. But it is also not uncommon for senior scientists to find their work challenged by Young Turks at some point in their careers, and to consider readjusting their positions. A combined failure of nerve and of cognitive flexibility is not uncommon among these scientists and seems to produce a noteworthy conservatism in science. Scientists have not been found unwilling to make embarrassing revisions in their work.

Of course, avoiding any serious challenge to established beliefs is not a problem of scientists alone. There is ample evidence that other groups (and probably all human beings) have difficulty incorporating ideas that conflict with their extant beliefs. See, for example, what Piaget has to say about the resistance to learning (accommodation) among small children. See Piaget 1963; Inhelder and Piaget 1964. This reluctance adds a new dimension to Kuhn's ideas about the use of paradigms in science. See Kuhn 1970.

20. Knorr 1986.

21. Compare to Garfinkel, Lynch, and Livingston 1981.

22. Knorr-Cetina 1981.

23. For the collective nature of reality formation in labs, see Knorr-Cetina 1981; Knorr-Cetina and M. Mulkay 1983, especially the review of the field of laboratory studies. For backstage as a practice area, see Goffman 1959.

As the excerpt in the text also indicates, photographs were frequently used in the early stages of vent research to make areas familiar to a group of researchers, not all of whom could go on each dive. Photographs taken during the dives were not made systematically and often did not have the details that would allow scientists to use them as data or even as simple ostensive models with which to capture and check their observations. So what were they for? To some extent they were like tourist photographs—at least in the sense that they captured major landmarks and emphasized the expected features of vent fields. They showed that the scientist who took them had actually been to the vent site, just as a picture of the Eiffel Tower indicates that a tourist has really been to Paris. They also began to give the wider scientific community an image of the sites. See Bourdieu et al. 1965.

Certain selected photographs started to appear in journal articles and grant proposals as "signs" standing for the areas. Scientists who had never been to the vents could use the pictures as ostensive models (or illustrations) to get a sense of what the vents generally looked like and what differences there were between areas named "Clam Acres" and "Rose Garden." Researchers who had been to the vents sometimes also used the images as icons, embodying and conveying the symbolic capture of these areas by science. Their iconic purpose was revealed most strikingly by the fact that many of the photographs were reused in a variety of articles on the vents, particularly articles in popular magazines that bragged about the discovery to the public. They were used there less as ostensive models than as trophies. See Mukerji, "Contentious Images," paper prepared for the ISA meetings, New Delhi, 1986; and Mukerji 1985.

24. For the anthropological idea of thinking with objects, see Douglas and Isherwood 1979.

25. Galileo describes the book of nature as written in the language of mathematics in *The Assayer* [1623] 1964. See also Latour 1986.

26. We can see here the mechanisms by which paradigms works. See Kuhn 1970.

27. It is important to note here the role of mathematics in scientific debate, for the process of extracting science from the natural world. Scientists routinely contend that they cannot make "science" from their data unless they derive numbers from the data that will advance an argument. Knowing what argument you want to make is part of the problem; having the numbers to make it legitimate is another. In part, what they are describing is a formal aspect of scientific writing. Articles are supposed to be based on numerical data or models of some sort. But there is more to it than this. The translation of nature into mathematics is a central part of the alienation process. Mathematical models are the ideal models in science because they can be manipulated and applied to new data as it is worked up. Thus "having good numbers" is an important element of doing good science, helping to characterize the culture of science to be served by research. See Latour 1986. See also Cohen 1982 for the historical development of numeracy and its legitimating value in America.

CHAPTER NINE
DIRECTING SCIENTIFIC DISCOURSE

1. The basis for this kind of thinking in sociology of the professions is mentioned in Starr 1982, in the introduction. The application of these ideas to science can be seen in Barnes and Edge 1982, particularly the introduction. This idea is demonstrated more than discussed in Gilbert and Mulkay 1984.

2. Bourdieu 1975.

3. See Latour 1986; Bourdieu 1982. For examples of these practices, see Gilbert and Mulkay 1984 and Knorr 1987.

4. For mention of conceptual interdependence, see Pickering 1982; for the concept of tacit knowledge, see Collins 1982; for analyses of stakes in controversies see, Gilbert and Mulkay 1984, and Pickering 1982; for discussions of reputation as a reward system in science, see, for example, Merton 1973, 1970; Zuckerman 1977; J. Cole and S. Cole 1973.

5. Gilbert and Mulkay 1984; Latour 1984; Latour 1986; Zuckerman 1977; Watson 1968; Pickering 1982.

6. Gilbert and Mulkay 1984; Latour 1984; Latour 1986; Barnes and Edge 1982, introduction; Stewart and Feder 1987; Nelkin 1984.

7. Merton 1970; Thomas 1983. For the opposite point of view, see Leiss 1972.

8. Gilbert and Mulkay 1974; Latour 1984; Pickering 1982.

9. In spite of these cases of territoriality, with all their lack of detachment, the culture of dispassion that Keith Thomas describes has not entirely disappeared. Scientists and philosophers in the twentieth century may have cast doubt upon the ability of researchers to achieve any real detachment, but few others than Paul Feyerabend (1975) have advocated abandoning scientific methods and mak-

ing passion and intuition more respectable in the laboratory. Most observers now agree that science does not uncover truths about nature as much as provide tentative models of it. They just hope and expect that the models will move closer and closer to reality. But the realization that today's scientific "knowledge" is tomorrow's rejected or massively revised theory has not altered the dispassionate stance that researchers are supposed to aspire to during the research process. Now the goal may be accurate information rather than truth, but the road to it resembles the earlier one.

So how can one reconcile the efforts at formal detachment with the informal passion that abounds in science? One does not really have to. The formal system and informal one are invoked in quite disparate moments in the research situation. And researchers make at least some routine attempts to keep them separate. A measurement will be discarded if it is botched by a fit of anger with the intransigence of the materials. In this and similar ways, scientists try to purify the research process in some obscurely understood but widely shared fashion. Like the reporters that Dan Hallin studied, they do this by making fine grain distinctions among situations and what they should do to maintain their objectivity. The obvious result is a system of social ritual, used for purification.

Whether or not these purification rituals do the job (and scientists themselves can never really be sure how uncontaminated their results really are), they at least display an allegiance to the spirit of detachment in relation to nature that is thought to be at the heart of science. And this normative standard holds in check (at least to some extent) the passions behind the territoriality in science. Compare to Schudson 1978 on the news.

10. Simmel 1950a; Coser 1956. This idea is also essential to the conception of class consciousness in Marx and some other Marxists. A class only exists when it is made self-conscious through its differentiation from others. See this idea reworked in Thompson 1963. See it disputed in Jones 1983.

11. Hughes 1971.

12. Latour 1986; Pickering 1982; Collins 1985.

13. Compare to Pickering 1982, Collins 1985, Latour 1986. For discussion of honorary authorship in science, see Stewart and Feder 1987, particularly p. 210. See also Braunwald 1987.

14. Stewart and Feder 1987; Braunwald 1987.

15. Compare to Baxandall's description of how Picasso managed his reputation by choosing when to be identified with a group, and when to show independence. See Baxandall 1985, chapter 2.

16. For some of the problems and subtleties of the "ownership" of scientific findings, see Nelkin 1984.

17. Corliss et al. 1979; Grassle 1982; Spiess 1980; Hayman and Macdonald 1985; Jannasch and Mottl 1985; Edmund et al. 1982.

18. Stewart and Feder 1987; Braunwald 1987.

19. These ideas would make problematic some ideas in Merton 1970, 1973; Zuckerman 1977; and others.

20. See Latour 1986 on "black boxing."

21. Like Pasteur. See Latour 1984.

CHAPTER TEN
THE VOICE OF SCIENCE

1. See Habermas 1970 and 1981. The autonomy of scientists remains such a fundamental issue in the world of science in spite of its severe limits partly because the government has interests in supporting the image of science as a powerful and detached knowledge system. The state's relationship to science is governed by a need for useful advice—the need for experts, technocrats, to solve the problems that come (as Habermas points out) with the rationalization of political life in the modern state. Governments need science to be powerful because they gain some of their legitimacy by getting researchers to find cures for diseases or by inventing new communications systems or space vehicles that make their nations seem particularly clever or blessed with talent. On a more practical level, scientific experts enhance the power of the state through the kinds of breakthroughs in science or technology that can strengthen the military or the economy. For these reasons, science must seem powerful. It also must seem detached if scientific testimony is going to be at all persuasive in shaping government policies. The value of science in policymaking comes from its objective treatment. Perhaps most importantly, if scientists are to perform their political functions vis-à-vis the government, if they are to believe in the objectivity of their policy assessments, and if they are to identify their successes with national pride, they must be treated as though they have some power and autonomy. Since they have (for the most part) little power, they are encouraged to celebrate in their minimal autonomy and not think about their power.

2. For ways the modern state identifies problems as technical ones needed expert solutions, see Habermas 1970, chapters 4–6.

3. Collins 1985.

4. See Gilbert and Mulkay 1984.

5. See Knorr 1986.

6. For a discussion of the guardedness of scientists vis-à-vis outsiders, see Barnes 1985.

7. See Callon 1986 on scallops; Star 1988; Henderson 1987; and Joan Fujimura, "Where Social Worlds Meet: Standardized Interfaces and Bandwagons in Science," to appear in *Social Problems*.

8. See Primack and von Hippel 1974.

9. See Primack and von Hippel 1974.

10. Callon 1986.

11. See Morain and Meyer 1987; and Sheer 1988. In these articles, Roy Woodruff, who had been associate director for defense systems at Lawrence Livermore National Laboratory, is cited as saying that Edward Teller and a colleague, Lowell Wood, had knowingly made "overly optimistic, technically incorrect" estimates of the state of research on the X-ray lasers required by the Star Wars program. The point of his charge is that the scientists were willing to misrepresent the facts in order to encourage this line of research.

12. See Mehan and Wills 1989; Skelly 1988; Wertsch 1987.

13. Morain and Meyer 1987 and Sheer 1988 report that Teller had given the government inaccurately optimistic estimates of the possibilities for the X-ray la-

sers required for SDI, which provides an interesting case in point. The scientist who made this accusation (Roy Woodruff) was described as having been demoted for opposing the optimistic depiction of the laser research. In this case, Woodruff's demotion, which may have been designed to make him more willing to accept Teller's point of view, only made him feel that he had very little to lose by going public with the information.

14. See Primack and von Hippel 1974. The silencing function of unheeded research was illustrated nicely by the scientists who studied deep ocean disposal of nuclear waste. Their research finally recommended deep ocean dumping, but the virtues of this policy were never presented carefully to the public. Some of the scientists said that DOE officials never seriously considered ocean disposal. They had always advocated land-based disposal in spite of the threats to drinking water that this posed. By financing research on this alternative to land-based disposal, they could claim that they had looked into alternatives, but they had never really thought seriously about them. They claimed that the decision was a political one, in the sense that public opposition to ocean disposal was too great to overcome. This may well have been the case. But clearly the scientists who did this work were not given power because they were insiders to this policymaking. They were made into icons of careful policy evaluation, testaments to government commitment to detached analysis of possibilities. Then politics provided its own answer to the issue.

15. This idea bears a strong resemblance to the ideas presented by Bakhtin on heteroglossia as a kind of double-voiced discourse. See "Discourse in the Novel" in Bakhtin 1981, particularly p. 324.

16. See Habermas 1970 and 1984; Gouldner 1976.

17. Latour 1986.

18. This aspect of science has been pointed out by Latour and Woolgar 1979. They connect it to Eisenstein's arguments about print and the scientific revolution. Eisenstein sees the growth of modern science as contingent on the ability to put scientific thoughts and observations in print. Similar arguments are made about the importance of printed pictures and print in Ivins 1953.

19. See Wertsch 1987.

Bibliography

Alexander, C. M. 1975a. "Seafloor Effectiveness of RUM II." SIO Ref. Series 75-20, June 25, 1975.

Alexander, C. M. 1975b. "Sea Floor Technology Report No. 5, Sea Floor Effectiveness of RUM II." *Marine Technology Journal* 9:5–15.

Allmendinger, Eugene. 1982. "Submersibles—Past, Present, Future." *Oceanus* 25;18–29.

Anderson, V. C., and R. E. Todd. 1960. "Summary of an Engineering Design of a Remote Underwater Manipulator." SIO Ref. Series 60-11, Artemis Report No. 6, March 1960.

Bakhtin, M. 1981. *The Dialogic Imagination.* Austin: University of Texas Press.

Ballard, R. 1975. "Dive into the Great Rift." *National Geographic* 147: 604–15.

Ballard, R. 1982. "ARGO and JASON." *Oceanus* 25:30–35.

Ballard, R.D., et al. 1975. "Manned Submersible Observations in the FAMOUS Area Mid-Atlantic Ridge." *Science* 190:103–8.

Barnes, Barry. 1977. *Interests and the Growth of Knowledge.* London: Routledge and Kegan Paul.

Barnes, Barry. 1983. "On the Conventional Character of Knowledge and Cognition." In K. Knorr-Cetina and M. Mulkay, eds., *Science Observed.* Beverly Hills, Calif.: Sage.

Barnes, Barry. 1985. *About Science.* New York: Basil Blackwell.

Barnes, Barry, and David Edge, eds. 1982. *Science in Context.* Cambridge, Mass.: MIT Press.

Barnes, Barry, and Steven Shapin. 1979. *Natural Order.* London: Sage.

Baxandall, Michael. 1985. *Patterns of Intention.* New Haven: Yale University Press.

Baxter, James. 1968. *Scientists Against Time.* Cambridge, Mass.: MIT Press.

Beach, Mark. 1972. "Was There a Scientific Lazzaroni?" In George Daniels, ed., *Nineteenth-Century American Science*, Evanston, Ill.: Northwestern University Press.

Becker, Howard. 1982. *Artworlds.* Berkeley: University of California Press.

Bell, Daniel. 1974. *The Coming of Post-Industrial Society.* New York: Basic Books.

Bellaiche, G., et al. 1974. "Inner Floor of the Rift Valley: First Submersible Study." *Nature* 250:558–60.

Berger, Bennett. 1981. *The Survival of a Counterculture.* Berkeley: University of California Press.

Bermingham, Ann. 1986. *Landscape and Ideology.* Berkeley: University of California Press.

Boffey, Phillip. 1975. *The Brain Bank of America.* New York: McGraw-Hill.

Bourdieu, P. 1975. "The Specificity of the Scientific Field and the Social Conditions of the Progress of Reason." *Social Science Information* 14:19–47.

Bourdieu, P. 1982. *Leçon sur la leçon*. Paris: Editions de Minuit.

Bourdieu, P. 1984. *Distinction: A Social Critique of the Judgment of Taste*. Trans. R. Nice. London: Routledge and Kegan Paul.

Bourdieu, P., and Jean-Claude Passeron. 1977. *Reproduction in Education, Society and Culture*. Trans. R. Nice. London and Beverly Hills, Calif.: Sage.

Bourdieu, P., L. Boltanski, R. Castel, and J.-C. Chamboredon. 1965. *Un Art Moyen*. Paris: Editions de Minuit.

Brady, Michael. 1982. "Computer Vision." *Artificial Intelligence* 19:7–16.

Braunwald, Eugene. 1987. "On Analyzing Scientific Fraud." *Nature* 325:215–16.

Bright, Charles. 1898. *Submarine Telegraphs*. London: Cosby Lockwood.

Broadus, James. 1987. "Seabed Materials." *Science* 235;853–59.

Brockway, Lucile. 1979. *Science and Colonial Expansion*. New York: Academic Press.

Bush, Vannevar. 1960. *Science: The Endless Frontier*. Reprint of 1945 ed. Washington, D.C.: National Science Foundation.

Callon, Michel. 1986. "Some Elements of a Sociology of Translation. Domestication of the Scallops and the Fishermen of St. Brieuc Bay." in J. Law, ed., *Power, Action, and Belief*. Sociological Review Monograph. London: Routledge and Kegan Paul.

Carter, Samuel. 1968. *Cyrus Field: Man of Two Worlds*. New York: Putnam.

Childress, J. J., A. J. Arp, and C. R. Fisher. 1984. "Metabolic and Blood Characteristics of the Hydrothermal Vent Tube-Worm *Riftia pachyptila*." *Marine Biology* 83;109–24.

Chomsky, Noam. 1969. *American Power and the New Mandarins*. New York: Vintage.

Clarke, Arthur C. 1959. *Voice Across the Sea*. New York: Harper and Row.

Cleveland, W. S., and R. McGill. 1985. "Graphical Perception and Graphical Methods for Analyzing Scientific Data." *Science* 229:828–33.

Cohen, Patricia C. 1982. *A Calculating People*. Chicago: University of Chicago Press.

Cole, Jonathan, and Stephen Cole. 1973. *Social Stratification in Science*. Chicago: University of Chicago Press.

Cole, Leonard. 1983. *Politics and the Restraint of Science*. Totowa, N.J.: Rowman and Allanheld.

Cole, Stephen, Leonard Rubin, and Jonathan Cole. 1978. *Peer Review in the National Science Foundation, Phase One of a Study*. Washington, D.C.: National Academy of Sciences.

Colin, Norman, and Eliot Marshall. 1986. "Over a (Pork) Barrel: The Senate Rejects Peer Review." *Science* 234:145–46.

Collins, H. M. 1982. "Tacit Knowledge and Scientific Networks." In B. Barnes and D. Edge, eds., *Science in Context*. Cambridge, Mass.: MIT Press.

Collins, H. M. 1985. *Changing Order*. Beverly Hills, Calif.: Sage.

Corliss, J., et al. 1979. "Submarine Thermal Springs on the Galapagos Rift." *Science* 203:1073–83.

Coser, Lewis. 1956. *The Functions of Social Conflict*. Glencoe, Ill.: Free Press.

Crawford, Mark. 1985a. "Low-Level Waste Deadline Looms." *Science* 229:448–49.

Crawford, Mark. 1985b. "DOE, States Reheat Nuclear Waste Debate." *Science* 230:150–51.

Cromie, William. 1980. "The New Oceanography." *Mosaic* 11:2–7.

Cullen, Vicky, ed. 1983. "The University-National Oceanographic Laboratory System: The Research Fleet." Washington, D.C.: National Science Foundation.

Daniels, George. 1971. *Science in American Society.* New York: Knopf.

Daniels, George, ed. 1972. *Nineteenth-Century American Science.* Evanston, Ill.: Northwestern University Press.

Day, Deborah. 1985. *Guide to the Roger Randall Revelle Papers, 1928–1979 in the Archives of the Scripps Institution of Oceanography.* SIO Ref. Series 85-26.

Deacon, Margaret. 1971. *Scientists and the Sea, 1650–1900.* London: Academic Press.

DeGeorge, R., and F. DeGeorge. 1972. *The Structuralists from Marx to Levi-Strauss.* Garden City, N.Y.: Doubleday Anchor.

Dickson, David. 1984. *The New Politics of Science.* New York: Pantheon.

Douglas, Mary, and B. Isherwood. 1979. *The World of Goods.* New York: Basic Books.

Dupree, A. Hunter. 1972. "The Measuring Behavior of Americans." In George Daniels, ed., *Nineteenth-Century American Science.* Evanston, Ill.: Northwestern University Press.

Dupree, A. Hunter. 1986. *Science in the Federal Government.* Baltimore: Johns Hopkins University Press.

Eagleton, T. 1979. *Literary Theory: An Introduction.* Minneapolis: University of Minnesota Press.

Edgerton, Harold, and Gleb Udinstev. 1973. "Rift Valley Observations by Camera and Pinger." *Deep Sea Research and Oceanographic Abstracts* 20:669–71.

Edmund, John, and Karen von Damm. 1983. "Hot Springs on the Ocean Floor." *Scientific American* 248:78–93.

Edmund, J. M., et al. 1982. "Chemistry of Hot Springs on the East Pacfic Rise and Their Effluent Dispersal. *Nature* 297: 187–91.

Elias, N. 1982. "Scientific Establishments." In N. Elias, H. Martins, and R. Whitley, eds., *Scientific Establishments and Hierarchies.* Dordrecht: D. Reidel.

Elias, N., H. Martins, and R. Whitley, eds. 1982. *Scientific Establishments and Hierarchies* Dordrecht: D. Reidel.

Elliot, J. H. 1970. *The Old World and the New.* Cambridge, Eng.: Cambridge University Press.

Ellul, Jacques. 1964. *The Technological Society.* Trans. J. Wilkinson. New York: Vintage.

Erzkovitz, Henry. 1984. "Entrepreneurial Scientists: The Structure of Science and Normative Change." Paper presented at the ASA meetings.

Ewing, M., J. R. Worzel, and A. C. Vine. 1979. "Early Development of Ocean

Bottom Photography at Woods Hole Oceanographic Institution and Lamont Geological Observatory." In John Hersey, ed., *Deep Sea Photography*. Baltimore: Johns Hopkins.

Feuer, Lewis. 1963. *The Scientific Intellectual*. New York: Basic Books.

Feyerabend, Paul. 1975. *Against Method*. Atlantic Highlands, N.J.: Humanities Press.

Field, Henry M. 1893. *The Story of the Atlantic Telegraph*. New York: Scribners.

Foucault, M. 1979. *Discipline and Punish*. New York: Vintage.

Foucault, M. 1979. "What Is an Author?" In J. Harari, ed., *Textual Strategies*. Ithaca, N.Y.: Cornell University Press.

Freidson, Eliot. 1986. *Professional Powers*. Chicago: University of Chicago Press.

Galileo. *The Assayer*. [1623] 1964. Trans. Stillman Drake. Quoted in Hugh Kearney, *Origins of the Scientific Revolution*, p. 126. London: Longman's.

Garfinkel, H., M. Lynch, and E. Livingston. 1981. "The Work of Discovering Science Construed with Materials from the Optically Discovered Pulsar." *Philosophy of Social Sciences* 11:131–58.

Gaston, Jerry. 1978. *The Reward System in British and American Science*. New York: Wiley.

Gilbert, G. Nigel, and Michael Mulkay. 1984. *Opening Pandora's Box*. Cambridge, Eng., and New York: Cambridge University Press.

Girouard, Mark. 1978. *Life in the English Country House*. New Haven: Yale University Press.

Girouard, Mark. 1979. *The Victorian Country House*. New Haven: Yale University Press.

Goffman, Erving. 1959. *Presentation of Self in Everyday Life*. Garden City, N.Y.: Doubleday.

Goffman, Erving. 1974. *Frame Analysis*. New York: Harper and Row.

Goffman, Erving. 1976. *Gender Advertisements*. Washington, D.C.: Society for the Anthropology of Visual Communication.

Golden, William T., ed. 1988. *Science and Technology Advice to the President, Congress, and Judiciary*. New York: Pergamon Press.

Goodell, Rae. 1975. *The Visible Scientists*. Boston: Little, Brown.

Goody, Jack. 1977. *Domestication of the Savage Mind*. Cambridge, Eng.: Cambridge University Press.

Gornick, Vivian. 1983. *Women in Science*. New York: Touchstone.

Gouldner, Alvin. 1976. *The Dialectic of Ideology and Technology*. New York: Seabury Press.

Grassle, J. F. 1982. "The Biology of Hydrothermal Vents." *Marine Tech. Soc. J.* 16:33–38.

Grassle, J. F. 1985. "Hydrothermal Vent Animals." *Science* 229:713–17.

Greenberg, Daniel. 1967. *The Politics of Pure Science*. New York: New American Library.

Gummett, P. 1980. *Scientists in Whitehall*. Manchester, Eng.: Manchester University Press.

Gusfield, Joseph. 1981. *The Culture of Public Problems*. Chicago: University of Chicago Press.

Habermas, Jürgen. 1970. *Toward a Rational Society*. Trans. J. Shapiro. Boston: Beacon Press.

Habermas. Jürgen. 1981. *The Theory of Communicative Action*, vol. 1: *Reason and Rationalization of Society*. Boston: Beacon Press.

Hagstrom, Warren. 1965. *The Scientific Community*. New York: Basic Books.

Hall, A. R. 1963. *From Galileo to Newton*. New York: Harper and Row.

Hallin, Daniel. 1986. *The "Uncensored War."* New York: Oxford University Press.

Hampton, Benjamin. 1970. *History of the American Film Industry from Its Beginnings to 1931*. Reprint of 1931 edition. New York. Dover.

Harari, J. 1979. *Textual Strategies*. Ithaca, N.Y.: Cornell University Press.

Hay, Cynthia. 1982. "Advice from a Scientific Establishment." In N. Elias, H. Martins, and R. Whitley, eds., *Scientific Establishments and Hierarchies*. Dordrecht: D. Reidel.

Hayman, R., and K. Macdonald. 1985. "The Geology of Deep-Sea Hot Springs." *American Scientist* 73:441–49.

Headrick, Daniel. 1981. *The Tools of Empire*. New York: Oxford University Press.

Heberlein, H. 1971–72. "Historical Development of Diving and Its Contribution to Marine Science and Research." Royal Society of Edinburgh, *Proceedings* (Section B, Biology) 72:283–96.

Heirtzler, J. R. 1974. "FAMOUS: A Plate Tectonics Study of the Genesis of the Lithosphere." *Geology* 2:273–378.

Heirtzler, J. R. 1975. "Project FAMOUS: Where the Earth Turns Inside Out." *National Geographic* 147:586–603.

Henderson, Kathryn. 1987. "Codification in Science." Unpublished paper.

Hersey, John. 1979. *Deep Sea Photography*. Baltimore: Johns Hopkins University Press.

Hessler, Robert, and Peter A. Jumars. 1977. "Abyssal Communities and Radioactive Waste Disposal." *Oceanus* 20:41–46.

Hessler, R., and W. Smithey. 1984. "The Distribution and Community Structure of Megafauna at the Galapagos Rift Hydrothermal Vents." In Peter Rona, K. Bostrom, L. Laubier, and K. L. Smith, eds., *Hydrothermal Processes at Seafloor Spreading Centers*. New York: Plenum.

Hessler, R. R., et al. 1978. "Scavenging Anthropods from the Floor of the Philippine Trench." *Deep Sea Research* 25:1030–47.

Hewlett, Richard, and Oscar Anderson. 1962. *The New World, 1939–1946: A History of the Atomic Energy Commission*, vol 1. University Park, Pa.: Pennsylvania State University Press.

Hewlett, Richard, and Francis Duncan. 1969. *Atomic Shield, 1947–1952: A History of the United States Atomic Energy Commission*, vol 2. University Park, Pa.: Pennsylvania State University Press.

Hiatt, Blanchard. 1980. "Sulfides Instead of Sunlight." *Mosaic* 11:15–21.

Ho, Ping-Li. 1962. *The Ladder of Success in Imperial China*. New York: Columbia University Press.

Hochschild, Arlie. 1983. *The Managed Heart*. Berkeley: University of California Press.

Holton, Gerald. 1973. *Thematic Origins of Scientific Thought*. Cambridge Mass.: Harvard University Press.

Hughes, Everett. 1971. *The Sociological Eye*. Chicago: Aldine.

Inhelder, Barbel, and Jean Piaget. 1964. *The Early Growth of Logic in the Child*. New York: Harper and Row.

Isaacs, J., and R. Schwartzlose. 1975. "Biological Applications of Underwater Photography." *Oceanus* 18:24–30.

Ivins, William. 1953. *Prints and Visual Communication*. Cambridge, Mass.: MIT Press.

Jannasch, H., and M. Mottl. 1985. "Geomicrobiology of Deep-Sea Hydrothermal Vents." *Science* 229:717–25.

Johnson, K., et al. 1986. "In Situ Measurements of Chemical Distributions in a Deep Sea Hydrothermal Vent Field." *Science* 231:1139–41.

Jones, Gareth Stedman. 1983. *Languages of Class*. Cambridge, Eng.: Cambridge University Press.

Keller, E. F. 1984. *Reflections on Gender and Science*. New Haven: Yale University Press.

Kevles, Daniel, 1978. *The Physicists*. New York: Knopf.

Knorr, Karin. 1986. "Visual Work: The Fixation of Evidence in Natural Science Inquiry." Colloquium in the Sociology Department, UCSD, January 8, 1986.

Knorr-Cetina, Karin. 1981. *The Manufacture of Knowledge*. New York: Pergamon.

Knorr-Cetina, Karin, and M. Mulkay, eds. 1983. *Science Observed*. Beverly Hills, Calif.: Sage.

Koshland, Daniel, Jr. 1985. "Modest Proposals for the Granting System." *Science* 229:231.

Kuhn, Thomas. 1970. *The Structure of Scientific Revolutions*, 2d ed. Chicago: University of Chicago Press.

Kulm, L. D., et al. 1986. "Oregon Subduction Zone." *Science* 231:561–66.

Kurzweil, E. 1980. *The Age of Structuralism*. New York: Columbia University Press.

Lakoff, Sanford. 1966. *Knowledge and Power*. New York: Free Press.

Latour, Bruno. 1986. *Science in Action*. Milton Keynes, U.K.: Open University Press.

Latour, Bruno, and Jocelyn De Noblet. 1985. "Les Vues de L'esprit." In *Culture Technique*, no. 14.

Latour, Bruno, and Steven Woolgar. 1979. *Laboratory Life*. Beverly Hills, Calif.: Sage.

Latour, Bruno, and Steven Woolgar. 1982. "The Cycle of Credibility." in B. Barnes and D. Edge, eds., *Science in Context*. Cambridge, Mass.: MIT Press.

Latour, Bruno. 1983. "Give Me a Laboratory and I Will Raise the World." In

K. Knorr-Cetina and M. Mulkay, eds., *Science Observed*. Beverly Hills, Calif.: Sage.

Latour, Bruno. 1984. *Les Microbes*. Paris: Metailie.

Latour, Bruno. 1986. *Science in Action*. Milton Keynes, U.K.: Open University Press.

Law, J., ed. 1986. *Power, Action, and Belief*. Sociological Review Monograph. London: Routledge and Kegan Paul.

Law, John. 1986. "On Power and Its Tactics: A View from the Sociology of Science." *The Sociological Review* 34:1–38.

Le Pichon, X. 1975. "Le Project FAMOUS: Pour Quoi Faire?" *La Recherche* 5:674.

Leiss, William. 1972. *The Domination of Nature*. New York: George Braziller.

Lupton, John, and Harmon Craig. 1981. "A Major Helium-3 Source at 15° S on the East Pacific Rise." *Science* 214:13–18.

Lynch, Michael. 1985a. *Art and Artifact of Laboratory Science*. London and Boston: Routledge and Kegan Paul.

Lynch, Michael. 1985b. "La rétine exteriorisée: Sélection et mathématization des documents visuels." *Culture Technique* 14:108–22.

McConnell, Anita. 1982. *No Sea Too Deep*. Bristol, U.K.: Adam Hilger Ltd.

Macdonald, Ken, and Bruce Luyendyk. 1981. "The Crest of the East Pacific Rise." *Scientific American* 244:100–16, esp. pp. 100–102.

McNeill, William. 1982. *Pursuit of Power*. Chicago: University of Chicago Press.

Mangone, Gerard. 1977. *Marine Policy for America*. Lexington, Mass.: Lexington Books.

Manheim, F. T. 1986. "Marine Cobalt Resources." *Science* 232:600–608.

Marr, David. 1982. *Vision*. New York: W. H. Freeman and Co.

Matkin, Joseph, letter to Mother Sarah Craxton Matkin, at sea St. Thomas on HMS *Challenger*, March 3, 1873. "Substance of Professor Wyville Thompson's Lecture, to the ship's company of H.M. Ship *Challenger*, on the Geography of the sea, & the object of the Challenger Expedition. With remarks on the Progress hitherto made."

Maxwell, A. E. 1970. *The Sea*. New York: Wiley.

Mehan, H., and John Wills. 1988. "MEND: A Nurturing Voice in the Nuclear Arms Debate." *Social Problems* 35:363–83.

Menard, H. W. 1986. *The Ocean of Truth*. Princeton, N.J.: Princeton University Press.

Merton, Robert. 1970. *Science, Technology and Society in Seventeenth-Century England*. 2d ed. New York: Howard Fertag.

Merton, Robert. 1973. *The Sociology of Science*. Chicago: University of Chicago Press.

Miller, Howard. 1972. "The Political Economy of Science." In G. Daniels, ed., *Nineteenth-Century American Science*. Evanston, Ill.: Northwestern University Press.

Milne, P. H. 1979. "Combined Hydrographic and Underwater Position-Fixing Systems." *Dock and Harbour Authority* 60:192–94.

Morain, Dan, and Richard E. Meyer. 1987. "Teller Gave Flawed Data on X-Ray Laser, Scientist Said." *Los Angeles Times*, October 21, 1987, part 1, pp. 1, 14.

Mukerji, C. 1981. *From Graven Images.* New York: Columbia University Press.

Mukerji, C. 1985. "Visualization and Cognition in Science." *Visual Communication* 10:30–44 and *Culture Technique* 14:208–23.

Mukerji, C. 1986. "Contentious Images." Paper prepared for the ISA meetings, New Delhi.

Mulkay, Michael. 1972. *The Social Process of Innovation.* London: Macmillan.

Mulkay, Michael. 1979. *Science and the Sociology of Knowledge.* London: George Allen and Unwin.

National Science Board. 1987. *Science and Engineering Indicators—1987.* Washington, D.C.: U.S. Government Printing Office.

National Science Foundation. 1980. *Mosaic* 11:2–28.

National Science Foundation. 1981–83. *Federal Funds for Research and Development, Fiscal Years 1981, 1982, 1983,* vol. 3: *Surveys of Science Resources Series.* NSF 82-326, table C-84.

National Science Foundation. 1989. *Budget Summary Fiscal Year 1989.*

Needham, Joseph. 1969. *The Grand Titration.* Toronto: University of Toronto Press.

Nelkin, Dorothy. 1984. *Science as Intellectual Property.* New York: Macmillan.

Norman, Colin, and Eliot Marshall. 1986. "Over a (Pork) Barrel: The Senate Rejects Peer Review." *Science* 234:145–46.

Office of Management and Budget, Executive Office of the President. 1988. *Special Analysis J. Research and Development.* Reprint of pages J-1 through J-33 from *Special Analysis, Budget of the United States Government, 1989.* Appendix table 2-3, "National Expenditures for R & D, by Source: 1960–85," February 1988.

O'Rand, Angela. 1986. "Knowledge Form and Scientific Community." In G. Böhme and N. Stehr, eds., *The Knowledge Society.* Boston: D. Reidel.

Paight, Daniel. 1987. "Private Autonomy, State Science, and the Scientists' Movement." Unpublished paper, La Jolla, Calif.

Phillips, J. D., et al. 1979. "A New Undersea Geological Survey Tool: ANGUS." *Deep Sea Research* 26:211–25.

Piaget, Jean. 1963. *The Origins of Intelligence in Children.* New York: Norton.

Pickering, A. 1982. "Interests and Analogies." In B. Barnes and D. Edge, eds., *Science in Context.* Cambridge, Mass.: MIT Press.

Piven, Francis Fox, and Richard Cloward. 1971. *Regulating the Poor: The Functions of Public Welfare.* New York: Vintage.

Press, Frank. 1986. "Perspective." *Science* 231:1351.

Price, Derek de Solla. 1986. *Little Science, Big Science—and Beyond.* New York: Columbia University Press.

Price, Don. 1965. *The Scientific Estate.* New York: Oxford University Press.

Primack, Joel, and Frank von Hippel. 1974. *Advice and Dissent.* New York: Basic Books.

Raitt, Helen, and Beatrice Mouton. 1967. *Scripps Institution of Oceanography.* San Diego: Ward Ritchie Press.

Reingold, Nathan, ed. 1979. *Science in Nineteenth-Century America.* New York: Octagon Books.

Richter, M. N. 1980. *Autonomy of Science.* Cambridge, Mass.: Schenkman.

Rona, Peter, K. Bostrom, L. Laubier, and K. L. Smith. 1984. *Hydrothermal Processes at Seafloor Spreading Centers.* New York: Plenum.

Rose, Hilary, and Steven Rose. 1969. *Science and Society.* London: Allen Lane.

Rowland, Henry A. 1979. "The Highest Aim of a Physicist." Reprinted in N. Reingold, ed., *Science in Nineteenth-Century America.* New York: Octagon Books.

Russell, Colin. 1983. *Science and Social Change, 1700–1900.* London: Macmillan.

Sarton, George. 1957. *Six Wings.* Bloomington: Indiana University Press.

Schlee, Susan. 1973. *The Edge of an Unfamiliar World.* New York: Dutton.

Schor, Elizabeth Noble. 1978. *Scripps Institution of Oceanography: Probing the Oceans, 1936–1970.* San Diego: Tofua Press.

Schudson, Michael. 1978. *Discovering the News.* New York: Basic Books.

Schulenberger, E. E., and R. R. Hessler. 1974. "Scavenging Abyssal Benthic Anthropods Trapped under Oligotrophic Central North Pacific Gyre Waters." *Marine Biology* 28:185–87.

Sears, Mary, and Daniel Merriman. 1980. *Oceanography: The Past.* New York, Heidelberg, and Berlin: Springer-Verlag.

Shapin, Steven. 1988. "Technicians in 17th-Century England." Colloquium in Sociology, UCSD, April 12, 1988.

Shapley, Deborah, and Rustum Roy. 1985. *Lost at the Frontier.* Philadelphia: ISI Press.

Sheer, R. 1988. "The Man Who Blew the Whistle on Star Wars," *Los Angeles Times Magazine,* July 17, 1988, pp. 6–14, 29–32.

Sherry, Michael. 1977. *Preparing for the Next War: American Plans for Postwar Defense, 1941–45.* New Haven and London: Yale University Press.

Simmel, Georg. 1950a. *The Sociology of Georg Simmel.* Trans. K. Wolff. Glencoe, Ill.: Free Press.

Simmel, Georg. 1950b. "The Stranger." In *The Sociology of Georg Simmel*, trans. K. Wolff. Glencoe, Ill.: Free Press.

Skelly, J. M. 1988. "Power/Knowledge: The Problems of Peace Research and the Peace Movement." In C. Alger and M. Stahl, eds., *A Just Peace Through Transformation.* Boulder, Colo.: Westview Press.

Smith, Alice Kimball. 1965. *A Peril and a Hope: The Scientists' Movement in America: 1945–47.* Chicago: University of Chicago Press.

Spencer, D. 1982. "Ocean Science and Ships." *Oceanus* 25:1–4.

Spiess, F. N. 1980. "East Pacific Rise." *Science* 207:1421–33.

Spiess, F. N. 1987. "Up Periscope." Donald L. McKernan Lecture in Marine Affairs, University of Washington, February 10, 1987.

Spiess, F. N., and J. D. Mundie. 1970. "Small-Scale Topographic and Magnetic Features." In A. E. Maxwell, ed., *The Sea*, vol. 4. New York: Wiley.

Spiess, F. N., and R. C. Tyce. 1973. "Marine Physical Laboratory Deep Tow Instrumentation System." SIO Ref. Series 73-4:37.

Spiess, F. N., et al. 1969. "Detailed Geophysical Studies on the Northern Hawaiian Arch Using Deeply Towed Instrument Package." *Marine Geology* 7:501–27.

Star, S. Leigh. 1988. "The Structure of Ill-Structured Solutions: Heterogeneous

Problem-Solving and Boundary Objects." *Proceedings* of 8th AAAI Workshop on Distributed Artificial Intelligence, Lake Arrowhead, Calif., 1988.

Starr, Paul. 1982. *The Social Transformation of American Medicine.* New York: Basic Books.

Sternberg, Robert. 1985. "Human Intelligence: The Model is the Message." *Science* 230:1111–18.

Stewart, Irvin. 1948. *Organizing Scientific Research for War: An Administrative History of the Office of Scientific Research and Development.* Boston: Little, Brown.

Stewart, W. W., and Ned Feder. 1987. "The Integrity of Scientific Literature." *Nature* 325:207–14.

Strong, Roy. 1979. *The Renaissance Garden in England.* London: Thames and Hudson.

Sudnow, David. 1978. *Ways of the Hand.* Cambridge, Mass.: Harvard University Press.

Thacker, Christopher. 1979. *The History of Gardens.* Berkeley: University of California Press.

Thomas, Keith. 1983. *Man and the Natural World.* New York: Pantheon.

Thompson, E. P. 1963. *The Making of the English Working Class.* New York: Vintage.

Turner, Ruth. 1977. "Wood, Mollusks, and Deep-Sea Food Chains." *Bulletin of the American Malacological Union* 7:13–19.

Vetter, Richard. 1970. *Growth and Support of Oceanography in the United States.* Washington, D.C.: National Academy of Sciences, NCR.

Vine, Al. 1982. "The Case for Submersibles." *Oceanus* 25:1–17.

Waldrop, Mitchell. 1980. "Hot Springs and Marine Chemistry." *Mosaic* 11:8–14.

Wallerstein, Immanuel. 1976. *The Modern World-System.* London: Academic Press.

Watson, James. 1968. *The Double Helix.* New York: Atheneum.

Weingart, P. 1982. "The Scientific Power Elite—A Chimera." In N. Elias, H. Martins, and R. Whitley, eds., *Scientific Establishments and Hierarchies.* Dordrecht: D. Reidel.

Wenk, Edward, Jr. 1972. *The Politics of the Ocean.* Seattle: University of Washington Press.

Wertsch, J. 1987. "Modes of Discourse in the Nuclear Arms Debate." Working Paper No. 8. Chicago Center for Psychosocial Studies.

Whitteridge, G. 1971. *William Harvey.* New York: Elsevier.

Williams, Rob, and John Law. 1980. "Beyond the Bounds of Credibility." *Fundamenta Scientiae* 1:295–315.

Wollen, Peter. 1969. *Signs and Meaning in the Cinema.* Bloomington: Indiana University Press.

York, Herb. 1987. *Making Weapons, Talking Peace.* New York: Basic Books.

Ziman, John. 1984. *An Introduction to Science Studies.* Cambridge, Eng.: Cambridge University Press.

Zuckerman, Harriet. 1977. *Scientific Elite: Nobel Laureates in the United States.* New York: Free Press.

Index

"Black box" concept of technology, 131

Blue water oceanography, 18; funding accessibility for, 95–96

"Boundary objects" in science, 193–94

British government, deep ocean research funding, 29

Budgetary constraints with soft-money funding, 87

Bureaucracy, research funding and, 57–58

Bush, Vannevar: flawed projects, 222n.20; funding vulnerability, 104; scientific funding proposal, 53–61; on status of technicians vs. scientists, 125–26; subservience of technology, 138–39, 144; surveillance system, 65

Business: equipment production and consumption, 120; scientific research and, 12, 213n.24; technological dependence on, 119–24

Career planning: autonomy in research and, 88–93; funding dependency and, 9–10; group alliances and, 178, 232n.15

Carpenter, William, 27–30

Challenger expedition (1872–75), 27–30; design as research vessel, 31; fishery studies, 35–37

Chemistry: strategic importance of, 39; technological innovations in, 123

"Circulation System of Bikini and Rongelap Lagoons, The," 49–50

Civilian science, government funding of, 55, 316n.39

Class structure in science: attribution of credit and, 183–84; disciplinary rivalry and, 171–73, 232n.10; in laboratories, 138; research engineering and, 138–42

Cold War: research goals during, 47–49; scientific autonomy and, 87–88

Color theory of quarks, 168

Committee on Submarine Detection by Sound, 39–40

Communication skills: funding as, 62; importance of in ocean research, 69–72, 220n.9; informal vs. formal communication, 219n.2; information management and, 208n.5; politics of, 198–203; scientific research and, 213n.24

"Community acceptance" of projects, 95; power of science and, 4–14, 205n.2

Competition among scientists: Bush plan, 216n.40; disciplinary rivalries, 170–75;

intradisciplinary rivalry, 175–79; rivalry among project heads, 181–83; signaturization, 132–35; territoriality, 167–70

Computers: business role in development of, 121; deep ocean research, 111–12

Computer scientists, funding dependency of, 17

Conant, James, 217n.41

Congress: money manipulation and, 100–101; ocean research funding, 32; political basis for scientific funding, 219n.49; soft-money funding policies, 207n.4

Consulting by scientists: combined with research, 51–52; prestige of, 220n.6; reducing vulnerability with, 10

Cousteau, Jacques, 150–51

Criticism of research: flawed projects, 83–84, 222n.19; policymaking and, 208n.6. *See also* Peer review

Cronkite, Walter, 183, 187–89

Cultural aspects of science, 205nn.2–3; flawed projects, 222n.19; funding, x–xi; natural world and, 146–47, 227nn.2–3; technology and human capacity, 126, 225n.5

Cultural capital, scientists as, 210n.12

Cumulative advantage concept, 136, 226n.22; funding equity and, 222n.16; limits of autonomy and, 222n.2

Darwinian theories, impact on scientific research, 22; confirmation by Roger Revelle, 51

Data collection: declassification and, 117; decontextualization, 148; systems, 110

Deacon, Margaret, 28–29

Decentralization of funding, funding strategies and, 93–98

Declassified materials, technological innovation and, 113

Decontextualization, specimen collection and, 159–60

Deep ocean research: dependency on military technology, 113–19; developmental funding, 94; family life and, 72; lighting problems during, 117–18; politico-economic value of, 14–15; popular image of, 66; population studies, 36–37; postwar demobilization, 45–52; as research model, 14–20; research routines vs. excitement of diving, 67–68; social interaction during, 69–72, 220n.9; technology